ADVENTURES IN ORDER AND CHAOS

ASTROPHYSICS AND SPACE SCIENCE LIBRARY

VOLUME 313

ADVENTURES IN ORDER AND CHAOS

A Scientific Autobiography

By

GEORGE CONTOPOULOS
Member of the Academy of Athens,
Research Centre of Astronomy,
Athens, Greece

KLUWER ACADEMIC PUBLISHERS
DORDRECHT / BOSTON / LONDON

A C.I.P. Catalogue record for this book is available from the Library of Congress.

ISBN 1-4020-3039-8 (HB)
ISBN 1-4020-3040-1 (e-book)

Published by Kluwer Academic Publishers,
P.O. Box 17, 3300 AA Dordrecht, The Netherlands.

Sold and distributed in North, Central and South America
by Kluwer Academic Publishers,
101 Philip Drive, Norwell, MA 02061, U.S.A.

In all other countries, sold and distributed
by Kluwer Academic Publishers,
P.O. Box 322, 3300 AH Dordrecht, The Netherlands.

Printed on acid-free paper

springeronline.com

Printed in the Netherlands.

Contents

Preface

For many years I was organizing a weekly seminar on dynamical astronomy, and I used to make some historical remarks on every subject, including some anecdotes from my contacts with many leading scientists over the years. I described also the development of various subjects and the emergence of new ideas in dynamical astronomy. Then several people prompted me to write down these remarks, which cannot be found in papers, or books. Thus, I decided to write this book, which contains my experiences over the years.

I hope that this book may be helpful to astronomy students all over the world. During my many years of teaching, as a visiting professor, in American Universities (1962-1994, Yale, Harvard, MIT, Cornell, Chicago, Maryland and Florida) I was impressed by the quality of my graduate students. Most of them were very bright, asking penetrating questions, and preparing their homework in a perfect way. In a few cases, instead of a final examination, I assigned to them some small research projects and they presented their results at the end of the course. They were excellent in preparing the appropriate slides and in presenting their results in a concise and clear way.

On the other hand my Greek students, and students of several European Universities, were shy, awkward with the English language, and had difficulties presenting their work. But all this changed in the last twenty years. At a doctoral school in Thessaloniki during 1993 (Contopoulos et al. 1994b) about 50 students from many European countries were asked to make short presentations of their doctoral research. This time their presentations were perfect. They had nothing to envy from their American colleagues.

The change was not easy. It required a major effort to overcome the difficulties of the language, to learn how to find and use the literature, and how to present their results in a clear way.

I remember an assistant of mine, who had difficulty in speaking English in our seminars (we used English for practice). But I insisted. And the outcome was that a few years later he was a professor, and an associate editor of an international journal of Astronomy.

Thus, I want to encourage students to try hard. Even if they come from a remote country, or region, they can excell in international science.

My scientific autobiography relates my own efforts to overcome the difficulties of my background and reach the most up-to-day problems of dynamical astronomy. In particular it describes how I became involved in many problems, from celestial mechanics to cosmology, and what were the main ideas that guided me over the years. Finally I describe my experience in writing and refereeing papers, and in preparing lectures, as well as my thoughts about the future of dynamical astronomy and of related fields.

1. YEARS OF STUDY

I was born in a relatively small Greek town on the northern coast of Peloponnese, called Aigion. I attended primary and secondary school there. At the same time I started studying French, German and English. My mother was a teacher of mathematics at a high school (she was the first Greek girl to study mathematics).

My father was a lawyer. But he had a broader education, and he was fond of reading popular books on science, especially in French. I remember that when I was 10 years old he tried to teach me relativity theory. Maybe I did not learn much at that time but I was impressed by the theories of physics and I started wanting to explore the Universe.

During my high school years our country was occupied by foreign troops. Greece had conducted a victorious campaign when it was attacked by the Italian dictator Mussolini in 1940. The invaders were defeated and Greek troops entered deeply into Albania during 1940-1941. But then Hitler came to rescue his partner. The armoured German forces attacked Greece from its back, the weakly protected border with Bulgaria. Thus, Greece was defeated and was occupied by German troops for about four years. During most of that time Italian and Bulgarian troops participated in the occupation forces.

School continued, but with many difficulties. Our Gymnasium, one of the best buildings of our town, was taken by the occupation forces. Classes were continued in churches and in private houses. An immediate consequence of the occupation was starvation. Many people died of hunger, especially in Athens. The salary of my mother practically vanished because of the enormous inflation (after the occupation one new drachma was set equivalent to 50 billion inflationary drachmas). And my father had to take care of six children and three elderly grandparents.

We survived only because we were lucky enough to own a small farm outside the town. This farm contained many fruit trees, orange, apple, cherry and olive trees. We cultivated the soil under the trees twice a year, first with corn, and then with maize and vegetables. We had all kinds of small animals, goats, lambs, and chicken.

We were going to school in the town (half an hour fast walking by foot every day) and then came back to work in the farm. After dark there was a permanent curfew imposed by the occupation forces. But we were obliged from time to time to go out to water the fields, because water was available only during the night. It was scary to do that, especially when we were hearing the patrols in the nearby streets, shooting at everything that moved. Sometimes we had also visits from the guerrillas that were coming from the nearby mountains and were taking whatever they could carry.

In the mountains a guerilla warfare was going on continuously. As the Germans had certain losses their reprisals were terrible. The worst incident in our neighborhood was the burning of Kalavrita (a town of 2000 people where the Greek revolution against the Turks had started in 1821) and of the nearby villages. All the male population of Kalavrita (1200 men) were executed and fires were set in all buildings, including the building of the primary school, where all women and children were imprisoned. Only the last moment the door was broken and thus women and children escaped death. These people then came down to our town as refugees, asking for food and shelter.

As I was speaking French and some German I was used by my family as an interpreter with the Italians and the Germans, whenever they were coming to our house. The Germans were pretending to "buy" whatever they were taking. They "paid" us in German money ("occupation marks"), but it was written on them that they were not valid in Germany!

When Italy capitulated, the Italian soldiers were taken prisoners by the Germans. However, many Italians escaped and were hiding. A few came to work in our farm. Then the Germans posted an announcement that any one supporting the Italians would be executed. After that my father said: "I sympathize with these people, but I do not want to be executed because of them" and he sent them away. But my mother sent me every day to bring them food, in a remote place.

Life in the farm was hard. During daytime we worked in the farm, and the evening hours were devoted to reading and preparing our homework. We had no electricity, but we were reading around a big table with an oil lamp at the center.

Then I made my first "discovery". I finished reading the regular books of my school, and I found, somewhere, the University books of my mother. They were books on differential and integral calculus, analytical geometry, etc. I asked my mother to help me, but she refused absolutely. In fact she was extremely busy, as she was taking care of the family. Thus, I decided to go on by myself. I even found extra time for reading. I had to take out the goats to feed in the boundaries of the farm. I was supposed to hold the cords of the goats by hand, not to allow them to enter the cultivated fields. Instead of that, I fastened the cords of the goats to my legs and I had my hands free to work with the books.

After the first difficulties with the new terminology (I remember that I had some difficulty in understanding the infinitesimals, because the book contained no clear definition), I went fast over these books. My brother, by two years younger, was wondering what I was reading. So I tried to explain a few things to him. I remember that I taught him the binomial formula without paper and pencil, as we were out in the fields, and he not only understood it, but he could, later, reproduce all the formulae on paper. He would become a talented mathematician but, instead, he became a lawyer.

After the German occupation we were for some time under the communist guerrillas but finally the Greek state was established and I was able to move to Athens and participate in the entrance examinations for the University.

I took two entrance examinations. One in mathematics and one in engineering. The second one was the most difficult, and the most prestigious. I was fortunate enough to pass, first among my colleagues, in both exams. Then I had to decide what to choose.

I liked mathematics and physics (including astronomy) very much. I wanted to study physics thoroughly, from elementary particles, up to the whole universe, and I realized (correctly as it turned out) that mathematics was the main tool for this study.

But everyone around me wanted me to study engineering. This was a prestigious career, with plenty of money, and glamour, at that time. I remember that I went to see my teachers in the high school and all of them congratulated me for entering the engineering school. When I told them that I wanted to study physics and mathematics they were astonished. They asked me: "What are you going to do as a career?" I replied that I should be a teacher of mathematics (I had not the slightest idea then of a research career in a University), but I should do research by myself. Then my director, who loved me as his son, said with indignation: "Curse you if you become a teacher like us".

Nevertheless I did not change my mind. I entered the school of mathematics and I never doubted that this was my destiny. And two years later (1948) I had my first reward. A docent at the University, Dr. D. Kotsakis (later professor of astronomy) asked me if I would like to work as an assistant (without pay, of course) at the department of astronomy. I accepted with enthusiasm. And that was the beginning of my career in astronomy.

The most interesting thing in the department of astronomy was its library. It was in a mess and very incomplete. But I put it in order and read many books on mathematics, physics and astronomy, without any guidance. In this way I learned a lot. Among other books I remember particularly Goursat's "Cours d' Analyse Mathematique" and Joos' "Lehrbuch der Theoretischen Physik". I still have detailed notes from these books.

Then, when I was still an undergraduate, at the 4th year of my studies (1950), I was appointed assistant. (Later on the possibility to be appointed as an undergraduate was eliminated by law). My salary was about $40 a month, just sufficient for survival. But it allowed me to devote myself only to my scientific work. My job was mainly to teach my fellow students spherical astronomy, astronomical instruments and celestial mechanics (including general relativity).

At the same time I started observing at the National Observatory of Athens. My first job there was to do "meridian observations". I soon realized that these observations were completely useless. The accuracy of our method (observing "by eye and ear") was ten times smaller than other current "impersonal" methods, so that even if this kind of observations was of any value, our results would be worthless.

Then I was engaged in other programs, variable stars and positions of minor planets. But we had some "minor" problems. E.g. we could not control the leakage of electricity, that would pass through our eyebrows as we were observing, instead of only illuminating the instruments. Even the dome and the shutter did not work properly. I had to climb on the dome and push to turn it to the proper position. In order to observe variable stars I spent one month at the astronomical station of Penteli, on a hill outside Athens. There was no road (only a narrow path up the hill), no electricity, no telephone, no running water (we brought water with a donkey), no cooked food (unless we walked down to a small restaurant in the nearest village), and the only bed I had was a portable army bed. Nevertheless I was happy to work there. But I could not find anything that should be called a real research project. Thus, I was deeply discouraged with observational astronomy.

Many years later, when I was in La Silla, Chile, I saw how my colleagues worked and I envied them. Sitting relaxed on a chair they had in front of them three computer windows, one to guide the telescope, another one to show the spectrum and a third one to do some preliminary reductions. Then I said that I could well be an observational astronomer under these conditions.

2. THESIS

As I was finishing my first degree, I had to find an appropriate subject for my thesis. I could not find an appropriate observational subject, thus I turned to theory. There was no theoretical tradition in Greek astronomy and I could not find any one to recommend an appropriate subject for me. But fortunately I found a paper by Bart Bok on globular clusters. Bok stated that the orbits of stars in globular clusters were not studied sufficiently up to now. I immediately liked this subject. I went to my professor of astronomy to propose him this subject for a thesis. There was only one professor of astronomy in Athens at that time, Dr. S. Plakidis. Dr. Plakidis was a devoted observational astronomer. He had worked with Eddington in Cambridge on the variations of the periods of variable stars. He was trying to modernize astronomy in Greece, by finding money to get instruments and telescopes. One major new telescope that we got around that time was the Newall 63 cm refractor that was donated to us by the University of Cambridge [1]. But we had to fight for every bit of money from

[1] During my first visit to Cambridge they showed me this telescope that was not used any more. I asked why. The Director replied that the weather and the seeing were not good anymore for observations. I said then that if we had this telescope in Greece, where the weather is much better, it would be used profitably, and the Director replied: "That is a good idea. If you want it write us a letter". Thus, I wrote immediately to professor Plakidis, and soon thereafter the telescope was donated to Greece.

the government and we improvised every construction with the most simple means. Professor Plakidis was saying "We rediscover astronomy from the Chaldean times to the present".

When I asked him to allow me to work on a theoretical subject he said that he could not help me. But if I would find another professor willing to supervise my thesis he would not object.

Thus, I went to the professor of mechanics. He liked my proposed thesis subject and promised to support me. But when I asked if he had any idea how I should proceed he said: "Oh! It is very simple. You know the 2-body problem. Add one further body and you will have the 3-body problem. Then a 4th body and so on until you build a cluster". But how should I solve (analytically at that time) the 3-, 4- etc. body problems? I am afraid that if I had followed his advice I would not have finished my thesis yet!

Therefore I tried a different method. I started by assuming a given spherical density distribution and considered orbits of test particles in it. This method worked perfectly. I found later that a particular case of this problem had been worked out already by E. Strömgren (1917). But fortunately I had already considered the most general case. I found that the orbits were rosettes, that filled in general a ring, with an azimuthal angle difference between maximum and minimum, which was between $\pi/2$ and π. I published my results (in German!) in Zeitschrift für Astrophysik (Contopoulos 1954) and my paper attracted some interest in later years (over 30 citations, although the German language eliminated many prospective readers).

The professor of mechanics helped me in another, very efficient way. By pushing me to work hard and finish my thesis as soon as possible. After my graduation I had to serve in the army. When I took my first leave of absence from the army, I came to Athens to my department. There I saw the professor of mechanics, who asked me how my thesis was progressing. "But, Sir, I said I am now in the army". "This is not an excuse", he said. Then I realized that I should find time to proceed with my thesis, even while I was serving in the army. And somehow I did find time, under the most strange conditions, sitting on my bed, or during classes, or by finishing the army work and continuing my thesis work at the office.

Then I had to do a numerical example. I had only a hand calculator, an elementary cobweb machine. In order to multiply by 23 one had to turn the handle 3 times, move one place to the left, and turn again the handle twice. That method was not fast but it worked. After I finished my service in the army, a first draft of the thesis was ready. I only had to type and retype revised versions three times to get a precious "final" copy for my professors, and an equally precious carbon copy for myself (there were no xerox machines at that time).

Thus, in 1952 my thesis was ready. According to the law I had to go to a printer and "publish" my thesis as a book. Then I had to pass the thesis' examinations, in front of the whole faculty. At that time there were no formal graduate courses and only the "published" thesis was essential.

But I, and some other assistants in physics and mathematics, felt that we had not yet a sufficient education for our further work. Thus, we decided to organize a kind of graduate courses, for a small group of people. These courses took place in the evenings, after finishing our regular work. We had courses on various topics, that proved very useful to us. I was giving a course on general relativity, other colleagues were giving courses on quantum mechanics, various advanced courses in mathematics, and so on.

At the same time I was reading books and papers like mad. As I had no one to guide me, I was reading both useful and useless books. I was interested particularly in stellar dynamics. In order to understand the dynamical problems I read a lot of celestial mechanics. No one was there to tell me that this was not fashionable, at least not for stellar dynamicists. That lack of advice proved very useful later, when I used the perturbation methods of celestial mechanics in stellar dynamics.

My second paper was a geometrical one. It dealt with the isophotes of ellipsoidal nebulae (Contopoulos 1956a). I noticed that the fact that the isophotes are concentric ellipses does not prove that these systems are oblate spheroids, as believed at that time. They could equally well be triaxial ellipsoids. In order to check that, I proposed to measure the radial velocities along the minor axes of elliptical galaxies. If the systems were spheroidal these velocities should be zero, but if they were triaxial they should be different from zero.

This paper attracted little attention at that time. But after 20 years some measurements of velocities along the minor axis were made, and these velocities were clearly nonzero. Thus, there were a lot of references to my paper (over 75) in later years.

My third paper (Contopoulos 1956b) dealt with orbits in elliptical galaxies. That was my first contact with the Astrophysical Journal. Another paper published in German (Contopoulos 1957a) dealt with the effects of the radiation pressure on the interstellar dust.

Then I read the formidable papers of Chandrasekhar in the Astrophysical Journal (1939, 1940), on galactic dynamics, published before his book "Principles of Stellar Dynamics" (1942), and I tried to generalize his results. I published a short paper on this subject, but I could not go very far. I only discovered a "principle": "Don't try to go beyond Chandrasekhar by following the same road. Your only hope is to try to find a different road" (I mentioned this "principle" at a meeting in honor of S. Chandrasekhar, several years later; Contopoulos, 1978a). The "new road" was the "third integral" (see section 5 below).

3. FIRST TRIPS ABROAD

The first opportunity to travel abroad was the 1955 General Assembly of the IAU in Dublin. Of course I had no support at that time, and I had to provide all the costs myself. But the benefit was enormous. On my way to Dublin I visited many observatories and astronomical institutes in France, England, Scotland, Ireland, Belgium, the Netherlands, Switzerland and Italy, and I had discussions with many astronomers.

Some discussions were particularly interesting. When I was at the University of London Observatory I described my current research to some colleagues. I was then studying the effects of the radiation pressure on interstellar dust, in forming concentrations, like “globules”. One colleague said: “Let us check if this is possible in principle”. Then he made an order of magnitude calculation and came with a number that was very small. Thus he concluded: “Your effect is completely insignificant”. I was not used in doing easily such order of magnitude calculations and I could not answer. But when I returned to my hotel I made carefully these calculations and found that my colleague had made a mistake. At some point there should be a multiplication by a factor 10^6 (one million). Instead, he had divided by the same number. Then I called this friend and told him that the effect was 10^{12} times larger, thus it was significant.

At the University of London Observatory I had another funny experience. I was in the library until late at night, and then I realized that all the staff had left and they had locked all the doors. What should I do? Finally I found in a telephone directory the home number of the Director, and I called him. He came fast to my rescue.

But I had also a few disappointments. I told one leading astronomer that I wanted to study the orbits of stars in galaxies in order to understand the spiral structure. His reply was that the spirals were probably due to gas in a magnetic field and not to stars. And he added “Why don’t you do observations as you have such a nice sky in Greece”? I was sorry to learn some time later that he asked one of his assistants to do exactly what I was doing, namely to calculate orbits in galaxies. After some years I reminded him of his words, and he replied “How could I guess that you would be able to do this work by yourself”?

I tried to learn how to use the primitive computers of that time to calculate orbits, but most people discouraged me. E.g. at the Greenwich Observatory they calculated the Nautical Almanac tables by using a complicated set of electric cords similar to those used in a hotel switchboard. But later that year (1955) I saw one of the first electronic computers (with bulbs) in Manchester. And one year later (1956) I had my first calculated orbits in Stockholm with the help of Per Olof Lindblad.

One of the most important results of my 1955 visit was an invitation by Prof. Bertil Lindblad to go to Stockholm in 1956.

I was bold enough to approach many of the leading astronomers in Dublin, Lindblad, Oort, Heckman, and many others, and discuss my work and my prospects.

Next year (1956) I visited Stockholm for one month. On my way to Stockholm I stopped in Bern, Switzerland, to meet Dr. R. Kurth, one of the few people working on stellar dynamics at that time. He had written a paper on orbits of stars in a globular cluster in the Astronomische Nachrichten (Kurth 1955), where he referred to my earlier paper on the same subject (Contopoulos 1954). In particular he made a favorable remark on my proof that the angle between the directions of pericentron and apocentron is between $\pi/2$ and π. This reference showed that my work did not pass unnoticed, and encouraged me to continue working in this direction.

Dr. Kurth wrote one year later an interesting book on the dynamics of stellar systems (Kurth 1957) where he described the three main approaches for the study of stellar systems (a) the n-body problem approach, (b) the continuum approach, and (c) the statistical approach. The first two parts of his book contain many interesting new results. However his attitude as regards statistical methods was quite negative. He noticed that such methods had not the required mathematical rigour. Nevertheless the greatest advances in dynamical astronomy were made exactly in this field, although a better mathematical foundation is still required.

After Switzerland I stopped at some Observatories in Germany, Denmark and Sweden and finally I reached Stockholm. Prof. Lindblad provided a room for me at the Saltsjöbaden Observatory, and the meals in a nearby school. We had dinner at 5 o'clock in the afternoon, something quite strange for me, used in the Greek custom of having dinner after 10 p.m. in the night.

Among other people I met there Mrs. Ivanowska from Torun, Poland. We had many talks on scientific questions but also on the current situation in Poland. She felt very strongly the lack of freedom under the communist regime. When I was leaving Sweden I wished her "a free Poland". She thanked me but she could not believe that a change was possible.

Prof. Lindblad suggested a subject of research for me, to generalize his theory of epicyclic orbits. That was not difficult. I extended his first order theory to all orders and the results were published in the Stockholms Observatoriums Annaler (1957b).

Lindblad liked my work, and I gave my first seminar on this subject. At the same time he discussed his own work with me. He gave me a draft of a paper and he asked my opinion. The next day I told him that I stopped already in page 1. I could not understand how he could go from Eq.(2) to Eq.(3). Lindblad replied smiling "But it is so simple! You perform a Fourier analysis and discard all the higher order terms beyond the first. Then if you discard also some other small terms you get Eq.(3)".

I could not imagine that so many assumptions were hidden between two lines. But gradually I learned the way of Lindblad's thinking. I was impressed by his emphasis on the gravitational aspects of spiral structure, and I tried to understand his ideas about density waves in galaxies, that preceded, by several years, the modern theories of spiral density waves. But, anyhow, his writing was really difficult to follow.

Some years later, in 1962, C. C. Lin told me that he wanted to develop a gravitational theory of spiral structure, including density waves. I told him that B. Lindblad had already developed such a theory. We went together to the MIT library and borrowed several volumes of the Stockholms Observatoriums Annaler. But a few days later Lin told me "I cannot make sense of these papers. I prefer to start from scratch, by myself".

This is one explanation why Lindblad's theories of spiral galaxies did not attract the attention they deserved. The other reason was his emphasis on leading spirals. It is ironic that, although he remarked, correctly, that the theory of density waves is true both for leading and trailing spirals, he went on to consider only leading spirals. Only in his last papers (1961,1963) he tried to develop a theory of trailing spirals, after the numerical experiments of his son, Per Olof Lindblad, who noticed that trailing arms were formed preferentially.

I was impressed by the numerical results of P.O. Lindblad, that dealt with the evolution of rings composed of a small number of "particles" on the plane of symmetry of a galaxy. At the same time I was trying to extend the epicyclic theory to the third dimension. But I could not proceed very far by analytical methods. Thus, I asked Per Olof Lindblad to help me by calculating with the help of a computer two orbits on the meridian plane of an axisymmetric galaxy.

These calculations (Contopoulos 1958, Fig. 1) were astonishing. I expected that the orbits would be ergodic, and they should fill the interior of the curve of zero velocity on the meridian plane. Instead of that, the orbits looked like Lissajous figures, but with curvilinear boundaries (curvilinear parallelograms as I called them) with their apices on the curve of zero velocity. These orbits were the cause of a theoretical study, which led later to the third integral of motion (section 5).

4. APPOINTMENT AS A PROFESSOR

In 1956 a vacancy was formed in the University of Thessaloniki. The previous professor, Dr. J. Xanthakis, was elected member of the Academy of Athens, and left Thessaloniki. The formation of such a vacancy was completely unexpected. In fact, there were only two chairs of astronomy in Greece at that time, one in Athens and one in Thessaloniki, and both were occupied by not too elderly professors. Thus I was surprised when two of my professors informed me of this vacancy and urged me to apply.

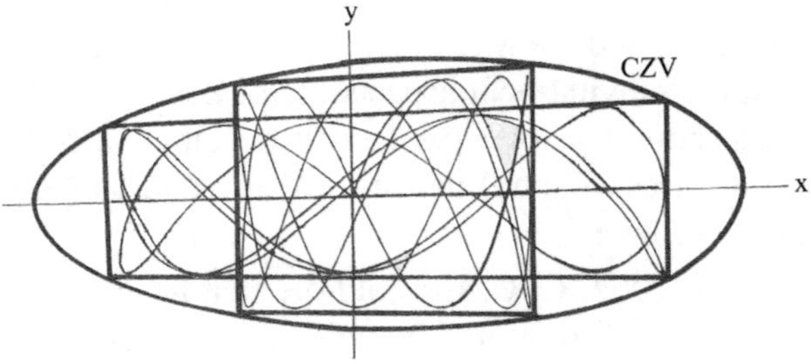

Figure 1. The first two orbits calculated in Stockholm (1956), inside the corresponding curve of zero velocity (CZV), filled two "curvilinear parallelograms".

According to the Greek standards at that time I was too young (then 28 years old) to apply for such a position. Meanwhile I had received, after some difficult exams, a scholarship to go as a postdoc to Chicago to work with Chandrasekhar. But I might not have the chance of applying for a similar position for many years to come.

When I visited the professors of the University of Thessaloniki they were initially reserved. But after several discussions with me, many of them changed in my favour. To the argument that I was too young, one professor replied: "But this drawback is reduced from day to day".

My work was obviously very limited at that time. But some people found that it might be worth to take the risk of appointing a young promising scientist. The election was not easy. Some professors wanted to postpone the election for three years. On the election day, after a long discussion, I received 8 favourable votes. Two voted for another candidate and three did not vote for any one. Furthermore two professors were absent and counted as negative votes.

I had the majority (8 out of 15 faculty members) but that was not sufficient. Unless I had a two thirds majority the minister of education had the right not to appoint me. According to the law there should be another final vote after one week. In this second meeting I received 10 votes. The result of this vote was now obligatory for the minister.

My election as a full professor was rather exceptional at that time. A full professorship was supposed to be a reward of a whole life of research. Thus, I understand the reservations of some faculty members.

After a few days I happened to meet my professor of physics at the University of Athens. He told me: "You are Mr. Contopoulos who was elected recently professor in Thessaloniki, right?" "Yes", I said. Then he gave me his hand. "I give you my condolences" he said. I was astonished. "Why?" And the professor said: "I considered you always as a promising young scientist. Now you are going to stop your research, and be absorbed by other duties. It is a pity". I protested vigorously. I said that I should continue now better my research work. But the professor was not convinced. Then I told him: "Let us have a bet. If after 20 years I will not have done what you expect from me I will accept your judgement". He agreed. After 20 years I was professor at the University of Athens, and some years later this same professor, then member of the Academy of Athens, was instrumental in my election as a member of the Academy.

When I arrived in Thessaloniki as a professor (1957) I realized that, despite the support of the faculty, my life would not be easy. I immediately took over heavy teaching duties (general astronomy, astrophysics, celestial mechanics, statistics, and general mathematics). Later on I took over also theoretical physics (I was the first to teach quantum mechanics in Thessaloniki) and various courses in mathematics.

The day of my arrival I learned that there was a computer committee, which was trying to get a computer for our University. I contacted them and we decided to ask the support of the University during the General Assembly of the professors a few days later.

At the General Assembly I described the need of a computer (then called an "electronic brain") to promote the research in our University. The older professors were looking at me with surprise and distrust. The rector asked: "Who else besides Dr. Contopoulos wants a computer?" No one supported me, not even the members of the computer committee. An old professor of humanities said: "My dear colleague, we, older professors, are used in using our own brains and we do not need the help of an electronic brain". Then the rector proposed: "Let us vote to support Dr. Contopoulos to travel abroad to do his calculations, but not buy a computer". This was voted unanimously with only me abstaining.

During the next year the request of Contopoulos was a current joke in the University. But the idea of a computer was not dead. After five years the University of Thessaloniki had the first modern computer in Greece (an IBM 1620). And then the young assistants and students started using the computer so completely, that I had great difficulty in finding computer time for my own research.

In Thessaloniki I had many good undergraduate students. Some of my best students decided to study astronomy. Among them were D. Kazanas (who was the first to propose the inflationary scenario for the early Universe), and B. Xanthopoulos (a leading relativist, and close collaborator of Chandrasekhar).

Later on many good students did their theses with me. During my 40 years of teaching I supervised over 40 theses. Most of these students had an exceptionally bright career. Over 30 became professors, and some others occupied high positions abroad.

At a meeting in Florida in my honor (1998) there were two former students of mine, Kazanas and Athanassoula. Kazanas (who did not do his thesis with me) presented two of my earliest papers, in Greek, and said that he was influenced by these papers to study astronomy. On the other hand Athanassoula represented my graduate students. She had done two theses with me, one in Greece and another one in France.

When I received a honorary Doctor's degree from the University of Chicago in 1991, the official citation referred to my "enthusiastic and generous encouragement of the work of younger colleagues". In fact I was happy to support many "younger colleagues" both in Greece and abroad (sections 48, 49).

When I arrived in Thessaloniki I was given only one room at the University for me and two assistants. Thus, my first efforts were directed in finding funds to build an Observatory in the University campus. My predecessor had already secured a 30cm telescope for this purpose. I had a friend who was an engineer and did the plans for the Observatory free of charge. Then I applied for money at the ministry. My request was turned down, but a few months later there were elections and a caretaker government was in charge for one month. I could approach the minister of finances and asked a small amount (about 7000 dollars at that time) to house the new telescope. The minister approved it. Then I brought him a written request for $10000. The minister was not happy. "This is more than you had asked for", he said, but in the end he agreed. Then I went to the University and, in a reform of the budget, we doubled this amount. We started building immediately. When the money was finished we went to the new minister for extra money. He told us: "What is this Observatory? According to my files such a building does not exist". "But it does" I said. Finally an extra amount was approved and I had to finish the building with the minimum possible cost. I had to negotiate every item personally with the constructors and thus I succeeded in getting prices far below the usual prices for the other buildings. The building was finished in a record time.

Then I had to find some good assistants. I had two new positions. One was taken by B. Barbanis, who did his thesis with me on the third integral. The other was taken by G. Bozis, who also did his thesis with me on the restricted three body problem. Both of them became professors later. B. Barbanis was an excellent collaborator, who helped me considerably and later became my successor.

The story with George Bozis is really funny. When he was appointed I was in a hospital in Athens with a broken leg. He visited me and asked what he was supposed to do. I gave him a list of books and papers that he should start reading. After one week I received a letter from him, saying: "I am sorry that I could not finish even one of the books that you gave me. Thus, I respectfully submit my resignation". I laughed and I only sent a message to him to wait until

I should be back in Thessaloniki. In fact G. Bozis was a very conscientious person. We did very good work together.

Another assistant of mine, that was appointed later, was J. Hadjidemetriou. He was one of my best assistants, and he had a very bright career. He is now accepted as one of the top people in celestial mechanics all over the world. He became not only professor, but also president of Commission 7 of the IAU, on celestial mechanics, a token of special recognition.

The salary of my assistants was very small. Thus, they had to subsidize their income by giving private lessons. In order to avoid that, I secured a research grant from the United States, which, I stated clearly, was to be used for compensation of my assistants, Hadjidemetriou and Bozis, to do research work beyond the usual working hours. My assistants were grateful and they worked with all their heart.

Many years later Hadjidemetriou was the Dean of the faculty of sciences in Thessaloniki, when I had moved to Athens. Whenever I visited my old department of astronomy in Thessaloniki, he came to pick me up from the airport and he was always driving me, whenever I moved in Thessaloniki. I told him that he should ask a student to do that for me, but he said: "I cannot do that to my professor".

I was not easy with my students, but I worked hard for them. Whenever my assistants came to me with a draft of a paper I made every effort to improve it. Thus, often the papers of my assistants were marked with too many red marks. But these papers could overcome many possible objections of the referees.

In one case an assistant of mine came with some results that were not really new. I told him that he should find similar results in Poincaré's "méthodes nouvelles de la méchanique céleste". "But I don't speak French" he said. "Then you should learn enough French to understand it" I said. "But the paper must be submitted to a meeting soon" he replied. "Not so soon" I said. "You have two weeks to learn French and one more week to read Poincaré".

Of course my suggestion was not realistic. But I wanted to emphasize the need to overcome the language barrier in dealing with scientific literature. In fact another assistant, Dr. C. Polymilis, learned enough French, German and Italian to be able to read scientific papers with the help of a dictionary.

5. THE THIRD INTEGRAL

In 1958 there was the General Assembly of the IAU in Moscow (Fig. 2). I had no support to travel, but I decided to go, paying all my expenses. I had to overcome all kinds of difficulties to get clearance to travel. My colleagues and friends tried to dissuade me."Are you not afraid?" they asked me. (At that time very few people from Greece had travelled to the Soviet Union).

It seems that my trip was considered suspect from both sides. In Moscow I was continuously escorted by an "interpreter", who was supposed to speak

Figure 2. The 1958 General Assembly of the IAU took place at the University of Moscow. Among the flags of the participating countries, there is a Greek flag (second).

Greek, but he knew only a few words, and never left me alone. On my departure, after I had put my notes from the General Assembly in my luggage, I realized later, when I was on the ship from Odessa to Pireus, that all my notes were stolen (not any valuable thing but my notes!). Then, when I returned to Greece, I was followed (not very discretely) for a long time by the Greek police.

At any rate I enjoyed my trip very much. I met many scientists, and had several useful discussions.

The most important discussion was when I presented the work I had done in Stockholm at a special meeting on stellar dynamics. When I showed my orbits on the meridian plane of a galaxy, a Soviet (Estonian) astronomer, G.G. Kuzmin, remarked that the form of the orbits indicated the existence of a third integral in this case. I replied that this was improbable, for two reasons (1) because of a theorem, due originally to Poincaré (1892), on the nonexistence of integrals beyond the energy (and the angular momentum in axisymmetric systems), and (2) because if such an integral existed in such a simple system[2] some one would have found it already. The conclusion of the discussion was that we should study this problem further.

A few months after my return to Greece I received a letter from Dr. Kuzmin, who said that he had looked at the theorem of Poincaré, and he was now convinced that there is no third integral.

This remark prompted me to look at the Poincaré theorem again. I realized that one condition for applying this theorem was not satisfied in my particular case[3].

Then, by using a series expansion, I could find a formal third integral explicitly, step by step, to all orders. Thus, I wrote to Dr. Kuzmin that *he* was right in Moscow, when he anticipated the existence of a third integral of motion.

I presented my new results at the Astronomische Gesellschaft meeting in Kiel in 1959. There was Prof. O. Heckmann from Hamburg, and he was immediately enthusiastic. My paper appeared in the Zeitschrift für Astrophysic in 1960 (Contopoulos 1960). This time my paper was in English; in fact, I had painfully realized that papers in German were not read by the majority of astronomers.

Strictly speaking the third integral is a formal (asymptotic) series, that converges only in exceptional cases. Although the original Poincaré theorem is not applicable to the simple potential (5.1), an extended form of this theorem is applicable, therefore one may conclude that there is no third integral as a function. But formal series, even if they are divergent, are very useful, if they are truncated at appropriate levels. Poincaré himself had made an extensive use of similar divergent series. Furthermore, in 1958-62 Kolmogorov, Arnold (a student of Kolmogorov) and Moser independently formulated the famous KAM theorem (after Kolmogorov, Arnold and Moser) that proves the existence of integral surfaces that fill most of phase space in systems like (5.1). The third integral represents these surfaces asymptotically.

---

[2] The galactic system I was considering referred to motions on the meridian plane (r, z) near the sun $(r = r_0)$. If we write $x = r - r_0$, y=z the potential is:

$$V = \frac{1}{2}(\omega_1^2 x^2 + \omega_2^2 y^2) - \varepsilon x y^2 - \varepsilon' x^3 \qquad (5.1)$$

plus higher order terms.

[3] This condition was that the Hessian of the unperturbed Hamiltonian H_0 should be different from zero. But in the present case:

$$H_0 = \frac{1}{2}(\dot{x}^2 + \dot{y}^2 + \omega_1^2 x^2 + \omega_2^2 y^2) \qquad (5.2)$$

and its Hessian is zero.

When I spoke to Moser about the third integral he was at first reserved, because of its formal character. But then he invited me to Courant Institute, in New York, to give a seminar on this topic. When he saw how well the third integral, truncated at higher orders, represented the integral surfaces, he was impressed and later he wrote: "These questions received renewed interest on account of the work of Contopoulos, who searched for and calculated a "third integral" for a system describing a galactic model" (Moser 1968).

A few years later a student of Arnold, Nekhoroshev (1977), developed a theory about the best truncation of the formal integrals, that is extensively used nowadays. In some cases the third integral has to be calculated to order 30, or higher, to give its best results. On the other hand the third integral is not applicable (not even approximately) in chaotic domains (sections 27, 35).

Therefore, my remark to Kuzmin in Moscow about the non existence of a third integral was wrong if by existence we mean a formal (and useful for applications) integral.

My second remark to Kuzmin turned out to be also wrong, because some people had already found some forms of the third integral. The first was Whittaker (1916, 1937), who had found an "adelphic integral" (from the Greek word adelphos, that means brother; brother to the energy integral). Whittaker expressed his adelphic integral in action-angle variables. Another formal integral was found in complex variables, by Birkhoff (1927) and by Cherry (1924). But Birkhoff never believed that the surfaces derived by the (formal) third integral were real surfaces. Instead, he believed that the orbits around an equilibrium point that is linearly stable (like the origin in Eq.(5.1)) can escape to large distances, and the origin is, in fact, unstable. It was only through the KAM theorem that the stability of the center in the system(5.1) was firmly established.

My form of the third integral was given in cartesian coordinates, appropriate for galactic dynamics. As it was shown later, the three forms of the third integral are equivalent. But only when higher order terms were calculated (by computer) the usefulness of the third integral was established. After that the third integral has been used extensively up to now, and several thousands of papers were devoted to it. In particular, there are more than 220 citations referring to my original paper in the Zeitschrift für Astrophysik, and over 2000 to my other papers on the third integral (the total number of citations to my work up to now is over 4500).

The calculation of higher order terms of the third integral is not an easy matter. In my first paper I calculated the terms of degree 3 and 4. The calculation of the 5th order terms (plus applications to the velocity ellipsoid of stars near the sun) was the subject of the thesis of B. Barbanis (1962), the first thesis under my supervision. But higher order terms could not be calculated by hand. Thus, I wrote a computer program to do the algebra and calculate the third integral to sufficiently high orders for many applications (Contopoulos and

Moutsoulas 1965). At that time there were no packages for computer algebra (like Mathematica today), thus I wrote the program in simple Fortran, for the IBM 1620 computer of our University. But then I noticed that the memory of the computer was not big enough to store the program and the results. Thus, I split the calculation in two parts. The first part gave the results in the form of punched cards (a deck one meter long). This deck was used as initial data for the second part that produced a deck 2 meters long. The final results were used then directly for applications, e.g. to calculate periodic orbits and invariant surfaces.

This work was published in 1965. It preceded a similar paper by Gustavson (1966) that is also much quoted nowadays. More efficient programs were developed by Giorgilli and Galgani (1978), Giorgilli (1979) and Efthymiopoulos et al. (2004), which are used extensively up to now.

6. TO THE UNITED STATES (THROUGH MOSCOW)

At the dynamical astronomy meeting in Moscow, where I spoke about my calculations in Stockholm, there was an elderly gentleman, who followed the discussion with much attention. After the meeting he came to me and introduced himself: "I am Dirk Brouwer" he said. "Would you like to come for some time to Yale?" I accepted gratefully and I promised to go there when I should have my first sabbatical.

In 1961 there was the IAU General Assembly in Berkeley. I got a grant from the IAU, but this only covered my travel to New York by plane. Thus, I travelled from New York to California by bus! This allowed me to see the country, and stop at some places to visit the astronomical institutes and observatories (Yale, Cleveland, Chicago, Los Angeles (Caltech and Palomar Observatory) and McDonald Observatory).

My most important stop was in Chicago. I visited the Yerkes Observatory, to see S. Chandrasekhar, and I stayed there for a few days. Chandra asked me to give a seminar about the third integral. After my talk, he summarized it with a few sentences. I was surprised how well he emphasized the main points. At the end he said: "Very good, George, that is very good". I replied: "Thank you Sir, you are very kind". His reaction was strange. He looked angry, and said: "I am not kind at all. And I will prove it to you". The next day there was a seminar by another astronomer, which was not particularly interesting. Chandra was quite rude in his remarks to him. At the end he came to me and said: "You see how kind I am?" During my stay in Yerkes I had long walks along Lake Geneva with Chandra. He talked to me about the future of dynamical astronomy. He said that there were two topics that should probably attract much interest in the coming years. One was the study of collective phenomena in galaxies and the other was general relativity. He was right in both areas. The study of collective phenomena materialized in the density wave theory, which was developed after a few years. And general relativity had an astonishing revival, due mainly to his own efforts. Then he asked me if I would like to work with him on general relativity. I was happy to accept, and I decided to split my sabbatical between Yale and Yerkes.

I continued my trip to California by the Greyhound bus, travelling day and night. On the way we stopped for dinner at a remote station along the highway. When I got up from my table I realized that my bus had left, taking all my luggage with it. What should I do? Someone suggested that I should take the next bus, after one hour. When I explained what happened to the driver of the second bus, he said: "Don't worry, we will catch it up". And by going fast we reached my bus after 5 or 6 hours!

In Berkeley I met several people that were doing work on stellar dynamics. Ivan King proposed to have an informal meeting at his department. There I met Donald Lynden-Bell, Michel Hénon and Sebastian von Hoerner among others. We were particularly impressed by the N-body calculations of S. von Hoerner, with 16 and 25 bodies. (Dr. W. Fricke, of Heidelberg, told me that he suggested to von Hoerner to do N-body numerical experiments after seeing my orbit calculations in Stockholm). We discussed the possibility of having a meeting on stellar dynamics. This idea led to the 1964 IAU Symposium No 25, in Thessaloniki, on the "Theory of Orbits in the Solar System and in Stellar Systems".

After the General Assembly most of us visited the Palomar Observatory (Fig. 3). Then I took the bus for the long trip to New York, stopping on the way at the McDonald Observatory in Texas. I had called a friend who was supposed to pick me at a particular junction on the highway. When the bus approached this point, I asked the driver to stop. The driver was surprised. "Are you sure that you want to get out here?" he said, "What will you do alone in the wilderness?" Fortunately my friend was there waiting for me.

7. YALE

Next year (1962-63) I had my sabbatical. Using also the summers of 1962 and 1963, I spent six months at Yale, five months at Yerkes, one month at Princeton (Institute for Advanced Study) and two months at the NASA Institute for Space Studies in New York.

When I arrived in New Haven (Yale University) I gave a series of lectures at a summer school on dynamical astronomy that was organized by Dirk Brouwer every year. I had asked him what they would expect from me, and he said: "The main thing is to give them appropriate references". Thus, I prepared extensive lists with all possible references, so that my style became a joke among the students. When another professor gave a reference, a student said: "This reference was not given by Contopoulos. Therefore it does not exist!".

Figure 3. With P.O. Lindblad at the Palomar Observatory (1961).

My lectures were later published as a chapter of the book "Space Mathematics I" (Contopoulos 1966a).

I remember that after one of my first lectures G. Clemence came to me and said: "During lectures I either sleep, or think about a recent problem of mine. But you forced me to follow your lecture and I could neither sleep, nor do anything else".

On one occasion a student asked details about a paper in German. I said that all he wanted was in that paper. "But I don't speak German" he said. "This will give you the opportunity to learn German", I replied [4].

[4]French and German have played an important role in the past. French has been for centuries the language of diplomacy, and German the language of philosophy. For this reason I encouraged my children to learn French as their first foreign language (they would anyhow learn English later). As regards German I had a number of very interesting philosophical discussions with German colleagues in a mixture of German and

Among the students there was a Greek student (G. Bozis) who had just arrived from Greece. In fact the first day of my stay at Yale I had asked Dr. Brouwer if he could support a student of mine to come there. Dr. Brouwer looked at his notebook. He wrote down his name and said: "Tell him to come next week". That was the blessed time, after the first Sputnik, when money was abundant in many US Universities.

Bozis had a problem. His English was rather poor. One example is the following. He had to pass exams in two languages. As he knew enough French he asked to be examined soon after his arrival. The professor then told him: "You passed well the French exams. But I doubt if you could pass an English exam". But he made great efforts to improve and after a few months his English was perfect.

As I write about languages I should mention a funny example. At a meeting someone was speaking in French. Next to me was C. C. Lin, who asked me: "Could you, please, translate for me?". "Of course" I said. Thus, I was listening carefully to the speaker and whispered to the ear of C. C. Lin my translation. But he looked at me with a strange face, and said: "But I do not speak Greek either". In fact, without realizing it, I was translating into Greek. By the way the translation to Greek was easy, almost automatic; it was much harder for me to mix two foreign languages and translate from French to English.

At Yale I made many friends. Besides Dirk Brouwer there were Morris Davis, Tony Danby, Harlan Smith, Victor Szebehely, Boris Garfinkel, Rupert Wildt, and many visitors.

In particular Morris Davis provided something like a family for me. I was very lonely at Yale. After the first days of welcome I was left alone. (Only the last week of my stay everyone realized that I was leaving and invited me to dinner every night). But Morris Davis invited me to his house once a week. He had five nice children. Every one was playing a musical instrument. And

---

English. The reason was that while I understand German I have difficulty in speaking it. On the other hand my colleagues could not express themselves clearly in English on philosophical matters. Thus they spoke in German and I replied in English.

English was established as the easiest and most broadly spoken language. Once established it was not easy to change things back. I remember one occasion, when I was giving a lecture in Paris, in French. I had difficulty in finding a French word for a "gap". I asked the audience: "How do you say "gap" in French?" And an astronomer replied: "gap"!

English, of course, can be very accurate and sophisticated, if one tries hard to learn it thoroughly. I remember Chandrasekhar, who was the editor of the Astrophysical Journal. He was very careful in using a correct language. On one occasion he was correcting in great detail a manuscript of mine, going to details, like punctuation, etc. Finally I protested:"But Chandra, I said, I think that my English is better than that of many Americans". And his reply was: "Who told you that all Americans know English?"

Another detail is the following: He was always using the word "algorism", instead of "algorithm". I thought that "algorithm" was an arabic expression for the Greek work "arithmos" (like "Almagest" is arabic for "Megisti mathimatiki Syntaxis" of Ptolemy). But Chandra insisted and brought me a dictionary, stating clearly that "algorithm" is wrong. In fact "algorism" is derived not from "arithmos" but from the name of a Persian mathematician Al Khowarismi.

when they sang they were excellent. In return I told them fairy tales. Thus, we became good friends, and I learn of their progress every Christmas even now.

I owe to Morris Davis also a lot as regards the use of computers. We had, first, a small computer, connected with a typewriter. When I wanted to calculate an orbit I was sitting next to the typewriter with a piece of graph paper. Every few minutes the typewriter would write the coordinates of a point along the orbit. I would mark the point on the paper and then wait for the next line. This process was incredibly slow. But even if I had all the computer results in front of me, the drawing of each orbit took several (3-5) hours.

At that time I was exploring the orbits in resonant Hamiltonians. In such cases the orbits are not deformed Lissajous figures, and I could not guess beforehand what they looked like. Only later, when I applied the resonant forms of the third integral, I could explain these orbits, that looked so strange at first.

A great progress was made a few months later, when the orbits could be drawn on paper by the computer. That increased my effectiveness by an order of magnitude at least.

When I returned to Greece, after my sabbatical, I tried to continue using the computers in the U.S.A. I was allowed to use the computer at the NASA Institute for Space Studies in New York. I had a friend there, to whom I was sending the initial data, in the form of a pack of cards. After about one month I was receiving the results, numbers and figures together. This slow method, by present standards, gave some unexpectedly good results.

A few years before that time two astronomers, Torgard and Ollongren (1960), had found numerically several types of tube orbits. These were so different from the box orbits (deformed parallelograms), that people doubted whether the third integral could explain them. But I always hoped that a resonant theory would explain these tube orbits, as due to particular resonances. Thus, when I developed the resonant theory completely, I could predict that such and such initial conditions would produce a particular type of tube orbits. I sent the initial data to New York. When the results arrived in the form of a long roll of paper, the whole staff of the department of astronomy came down to see the results. We unwinded the roll on the corridor. And behold! There appeared the nice tube orbits we expected (Fig. 4). We were so happy that we shouted with joy. Then we joined hands and danced a Greek dance around the tape, on the corridor.

My work on resonant forms of the third integral was published in four papers (Contopoulos 1963a, 1966b; Contopoulos and Moutsoulas 1965, 1966).

Many people ask what is the physical meaning of the third integral. In general we may say that in two degrees of freedom the energy can be split into two invariant parts, and the third integral is one of these parts. In the limiting case of two uncoupled oscillators, along the axes x and y, the energy of each oscillator is exactly conserved. Then the energy of only one oscillator (say

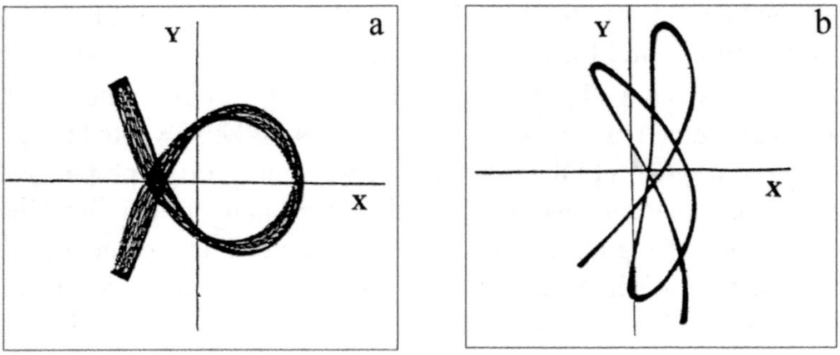

Figure 4. Resonant types of orbits, filling tubes of different shapes.

x) is the third integral. If there is a small coupling between the oscillators the energies of the oscillators vary a little. Therefore the third integral is an invariant part of the total energy, which, in the lowest approximation, is the energy of one oscillator. The coupling produces correction terms to the partial energy that are small if the coupling itself is small. In resonant cases the correction terms depend on the particular resonance. They are different for each resonance.

The most important resonances are 1:1 and 2:1. These are very different from all other resonances. In these cases the total energy cannot be split into two invariant parts and the third integral has a very different form; it is not the energy of one oscillator in the lowest order. Nevertheless the third integral exists also in these cases and it explains nicely the special forms of orbits.

When I was in Chicago one day I met Chandrasekhar in the corridor carrying a gadget to show to his students. It was an oscillator that could oscillate up and down, or rotate around a vertical axis. If it was set in up and down motion after some time the vertical oscillations stopped and all the energy was in rotational motion, i.e. oscillations leftwards and rightwards. Later the energy returned to the motion upwards and downwards. Chandra remarked to me that the partial energies of vertical and rotational oscillations were not preserved, therefore he thought that there was no "third" integral in the case. But I pointed out to him that this was a perfect example of a 1:1 resonant system, and in this case the third integral is very different from a partial energy.

At Yale, a most important event was the organization of a special Colloquium on modern dynamical problems (1962). We had among the participants, J. Moser, M. Hénon, A. Toomre, and others, most of them non-American. (Victor Szebehely made the remark that our American friends should feel embarrassed by being a minority).

Jürgen Moser (Fig. 5) spoke about the KAM theorem. But when I asked him if he could give a practical application in a simple astronomical problem, like the Earth-Moon system, he refused even to make a guess. Then Hénon worked out the inequalities of the proof of Moser and next day he gave us a number. This was $\mu \approx 10^{-48}$ (Hénon,1966) implying that the orbit of the Moon would be stable if the Earth and the Moon were point masses at a distance smaller than 1 mm. (A similar estimate for the Sun-Jupiter restricted 3-body problem would imply a mass of Jupiter, in its present position, of the order of the mass of a proton). But later the KAM theorem was improved to such a degree, that one now finds numbers that are of the same order as the numerically found limits of applicability. In particular an application to the stability of the Lagrangian points L_4 and L_5 in the Sun-Jupiter system gives results that include for the first time real asteroids in the stability region (see section 12.1).

Figure 5. Jürgen Moser at a meeting on "Space Mathematics" in Cornell University (1963).

While I was at Yale I had my first invitation to give a seminar in another University. I was invited by Malcolm Savedoff to go to the University of Rochester to speak about the third integral. The interest on this topic increased considerably over the years and I gave a large number of lectures about it all over the United States. A few years later I was invited again to Yale to speak about recent developments on this subject. Tony Danby, who presented me, said: "What one must do to become famous is to find an important problem and not solve it, but write many papers about it".

During my stay at Yale I wrote a number of papers. My first paper appeared in the Astronomical Journal (Contopoulos 1963b) and included a comparison of the work of Whittaker, Cherry and Birkhoff with mine, and more recent results. That paper includes a theorem, that convergent third integrals (not only formal) exist for potentials that agree with a given potential (in the form of a series) up to any given degree, and they differ only in higher order terms. Thus, integrable systems appear in every neighbourhood in a parameter space. But nonintegrable systems are even more numerous, like the irrational numbers in comparison with the rational numbers.

The original form of my Astronomical Journal paper included a discussion of the Poincaré maps on a surface of section. The paper was accepted for publication and it was in press, when the referee wrote me a belated "suggestion", to omit this part, as less important. As I had a great respect for referees in general, I followed his advice and eliminated this section from my paper at proof stage. That was a mistake. Soon afterwards appeared the well known paper by Hénon and Heiles (1964) and an important contribution of that paper is the then novel use of a Poincaré map on a surface of section.

Another development affected my assistant George Bozis. When I left Yale for Yerkes I suggested to him to try to find a third integral in the restricted three-body problem, by calculating invariant curves on a surface of section. But Victor Szebehely, the supervisor of Bozis, objected. He made the remark that the Poincaré theorem was applicable in the restricted problem and that proved that no third integral exists in this case. Bozis insisted that he had instructions from Contopoulos to do some numerical calculations to check this point. Finally Szebehely reluctantly gave him permission to use the computer for his calculations.

When Szebehely saw the results of Bozis he was amazed. He realized immediately that these results opened unexpectedly a new field of research. He got all these results and came to the IAU Symposium 25 in Thessaloniki, while Bozis stayed at Yale. During this Symposium Hénon presented his first results on the restricted three-body problem, showing the regions of order and chaos. At the discussion period Szebehely jumped up and presented the results of Bozis, almost a second paper on the subject.

I learned then a lesson. You cannot trust always the referee and the supervisor. You have to use your own judgement also.

8. CHANDRASEKHAR

I left Yale around Christmas 1962 for Chicago, and I arrived at Yerkes Observatory the 1st of January of 1963. What I saw surprised me. When I first visited Yerkes during the summer of 1961, it was a beautiful garden. A deep forest surrounded the Observatory and there were flowers everywhere. Not far away was Lake Geneva, full of boats, people swimming, and music. It was wonderful. Now the trees were denuded and deep snow covered everything. The ice of the lake was so thick that cars were circulating over it (I remember that in March I saw a notice "It is no more safe to drive on the lake").

The Observatory was quite isolated. During the week the students and the personnel were making the place alive. But during the week-ends no one was around. I found once a racoon in the dome of the big refractor (1 meter diameter), sitting on the observing chair. But the scientific atmosphere was excellent. I had many lively discussions, especially with Chandrasekhar.

When I arrived I asked him what we were going to do. He said: "I do not know. But let us read the book of Infeld and Plebanski, and we will find out". We found soon an interesting topic connected with the post-Newtonian approximations in general relativity. It was the virial theorem in the first post-Newtonian approximation. A joint paper was prepared soon (Chandrasekhar and Contopoulos 1963).

Then I continued with the post-Newtonian extension of the Lorentz formulae of special relativity. It is strange that no one had attacked this problem before. On the contrary some people assumed that a good generalization of the Lorentz formulae would be to replace u^2 in the Lorentz factor $(1 - u^2/c^2)^{1/2}$ by $(u^2 + 2V)$, where V is the potential energy, with the "justification" that u^2 is double the kinetic energy, and a generalization would be double the total energy. But this generalization is wrong. I developed the post-Newtonian approximation and I asked Chandra to publish a joint paper. At first he refused, but later he accepted and wrote me: "I am, in fact, embarrassed to call it joint (paper) since your part in it is almost the whole of it... I should feel less inclined to accept your generosity if I had not meanwhile independently arrived at the same result... Of course you went much further and clarified the problem with extraordinary completeness". Then, he wrote the paper in his own style and that paper attracted much interest in later years (Chandrasekhar and Contopoulos 1967).

During my stay at Yerkes (early 1963), Dr. Woltjer visited us for some days. We had many good discussions, but at first we disagreed about the third integral. Lo Woltjer thought that although the third integral was good for smooth potentials, it would not be applicable in more complicated cases like the spiral arms of a galaxy. Chandrasekhar was listening, and he tended to agree with Woltjer. Then Lo and I went to dinner together. I remember that I ordered fish and Lo ordered chicken. We started a very lively discussion, writing on the napkins of the restaurant. When we finished our meal I looked at the bones in our plates. Without realizing it, I had eaten the chicken and Woltjer had eaten the fish! But I had convinced Lo. We decided to write a paper together. Our paper appeared in the Astrophysical Journal (Contopoulos and Woltjer 1964) and dealt with the third integral in a rough model of spiral arms; that work preceded and was quite independent from the linear density wave theory of Lin, Shu, etc.

When we met Chandrasekhar the next day he said: "George, I thought about your problem and I am convinced that you are wrong and Lo is right". I replied: "Chandra, you are supposed to be an arbiter between Lo and me. But now Lo agrees with me, so you have also to agree".

On one occasion Chandra asked me what, in my opinion, was the importance of my work. I replied that it added to the understanding of some problems of stellar and galactic dynamics, and that was a small contribution in our understanding of Nature. That was rather prosaic, but Chandra was impressed favourably. He said: "I ask everyone around me the same question, and all feel offended. They reply that there is no doubt that their work is the most important in astronomy, and no one is above them in their field". Then I asked him in a slightly irreverent way: "And you, Chandra, do you find anyone who is above you?" He replied, becoming suddenly very serious: "Yes, I met two such people, Fermi and von Neumann". He had done research work with them, and he had a very high appreciation for both of them.

On another occasion we had dinner with Prof. B. Lindblad (a few months before his death). Chandra asked him the same question. And B. Lindblad replied: "There are several people above me, including the person who sits opposite to me" (Chandrasekhar).

There are many nice stories about Chandrasekhar. A phoney article of Chandrasekhar, that represented in a funny way his way of writing, was circulating among his students, and was finally published in the journal "Quarterly Journal of the Royal Society" (Candlestickmaker 1972).

Once we were discussing with him Nobel prizes and he told me the story of Yang and Lee, who discovered the parity nonconservation in elementary particles. Both were Chandra's students. He had a class once a week in Chicago with only these two students, and he was driving all the way from Yerkes for them (2 hours drive, and 2 more hours to return). Many people were surprised. "You are coming so far for only two students" they said. "But they are very good students" he replied. And they proved their value when they received the Nobel prize. Chandra would say: "My whole class received the Nobel prize".

Another story related with a Nobel prize was the following. Chandra attended all my lectures, while others (staff and students) did not come so regularly. I made this remark to him and he told me: "You know Cronin, who received the Nobel prize for the CP violation. He was attending my classes, which started early, at 8 o'clock in the morning. One day, as I was driving to my class, I heard an announcement on the radio that Cronin had received the Nobel prize the previous evening. Then I said to myself that Cronin would not come to class this day. But when I arrived, Cronin was there waiting. I went directly to him and congratulated him. The students were asking what happened. I told them that Cronin had just received the Nobel prize, and they clapped hands. At the same time, the press were trying to find Cronin, but they could not find him because he was attending my class".

In the following years Chandra and I became close friends. I used to visit him every time I was going to the U.S.A, and he visited me several times in Greece, with his lovely wife, Lalitha (Fig. 6).

Figure 6. S. Chandrasekhar and his wife, Lalitha, in their house in Chicago (1969).

I will relate another story from a later visit to Chicago (1969). Chandra had moved to Chicago and lived close to the University. I was giving a series of lectures on density waves, and Chandra attended all of them. Then people asked me to give a seminar at the Yerkes Observatory. When Chandra heard that, he said: "I will drive you there". I protested. "Chandra, I said, you know already the subject very well. I will only give a review there". "I want to hear this review" he insisted. So we went together. I had included in my slides a Table from a paper by C. C. Lin, where he compared the energies due to the various fields acting on a galaxy, like magnetic fields, cosmic rays and gravitational fields. But this Table was biased in favour of the gravitational fields, because it considered the whole gravitational energy, and not only the part due to the spiral arms. I was aware of this, and I wondered if I should present that slide at all. But as I started my talk I noticed that Chandra was soon asleep. Thus, I decided to show that Table, that was really spectacular. It indicated that gravitational effects were 1000 times stronger than magnetic and other effects.

But at the very moment I was emphasizing the gravitational effects, Chandra got up. "Wait a minute" he said. "That gravitational energy you have there contains all the background of the galactic rotation, is'nt it?" I tried to smile. "Chandra, you got me right on the act of the crime" I said. And I had to make some more calculations on the blackboard, showing finally that the gravitational energy due to the spirals was greater by a factor 10 than the magnetic energy. And Chandra said: "I can believe a factor of 10, but not 1000".

I return now to my first stay at Yerkes. Chandra wanted me to work with him, after the first post-Newtonian approximations, on the gravitational radiation in general relativity. Radiation terms appear in the 2 1/2 post-Newtonian approximation. I wanted to find the higher order post-Newtonian approximations by computer algebra, as I had done in the case of the third integral. But it was time to leave. When I came back to Chicago in 1969, Chandra had already done most of the basic work on gravitational radiation by himself.

I left Yerkes at the end of May. It was a pity to leave Yerkes at that time. The weather was beautiful, the trees were green again and full of flowers. But I had promised to Prof. Strömgren to go to the Institute of Advanced Study in Princeton.

Chandra was somewhat disappointed with my departure. As I was leaving he told me a funny story. Strömgren was then "the first citizen of Denmark" occupying a magnificent state owned mansion with a huge flower garden outside Copenhagen. The story referred to the previous "first citizen of Denmark", Prof. Niels Bohr. Bohr was invited to a party at the Royal Palace. When he entered the main room, the king was speaking to some foreign reporters about a recent soccer game, in which the goalkeeper, Mr. Bohr, had "saved the honour of Denmark at a critical moment". At that moment he turned towards Dr. Bohr. Bohr looked embarrassed. "Your Majesty, it was not me, but my brother", he said. Later on the king was speaking to another group and said the same words. Then Bohr said: "But I told your Majesty, that I was not in this game". And the king said angrily: "Cannot you pretend for one moment that you are your brother, who saved the honour of Denmark?".

9. PRINCETON - NEW YORK

When I arrived at the Institute for Advanced Study in Princeton, professor Strömgren was leaving. I was left alone to prepare a joint paper, or rather a book, published later by NASA, containing "Tables of Plane Galactic Orbits" (Contopoulos and Strömgren 1965). We adopted a new model of our Galaxy, trying to be as close to reality as possible. One reason for these calculations was to find the places of formation of the stars in the solar neighbourhood, knowing the ages of the stars. Some calculations were already done by a student of C. C. Lin, C. Yuan, and much more extensive calculations were done by P. Grosbøl, from Denmark, who did later his thesis with me in Greece. All these calculations indicated that most stars were born in spiral arms.

The Institute is located in the midst of a nice forest. Every day I was walking in the forest, and several times I could see deer grazing there.

I lived in a nice apartment at the limit of the forest, and I used the office of Dr. Strömgren when he left Princeton. I learned, later, that this office had been the office of Albert Einstein.

The Institute was ideally suited for work. It was very quiet, but at lunch time it was very lively. During that time I had discussions with many important people, that I knew only from the literature.

Among other people I met in Princeton Profs. Spitzer and Schwarzschild. Martin Schwarzschild was particularly inquisitive. He was very much interested in the third integral, and in later years he devoted himself to galactic dynamics. His most important contribution in this field was the construction of self-consistent galactic models. But he did also a lot of work on orbits and he advertised extensively the third integral.

After several years I was invited by Dr. Bahcall to give a lecture in Princeton. Martin Schwarzschild was also present. After my seminar we had a department lunch and Dr. Bahcall asked me to say a few words. I recalled my early discussions with Martin Schwarzschild and I concluded: "I was asked so many questions by Dr. Schwarzschild that although I work continuously on them for many years, I have not yet answered all of them".

The Institute was so isolated that one could not contact the outside world without a car. I realized that rather painfully one Sunday, when I had promised to be the best man in the marriage of a Greek friend of mine. The marriage would be in Perth Amboy, a town in New Jersey not very far from Princeton. I thought that it would be easy to find a taxi to get there. But when I tried to call a taxi there was no answer. I tried to walk along the road to downtown Princeton, looking for a passing-by taxi. Fortunately after some time a kind driver stopped near me and asked: "Do you need a ride?" "Oh yes", I said gratefully and I opened the door so abruptly that I had a good cut in my forehead. I was bleeding when I reached downtown Princeton and then Perth Amboy with a taxi. I was half an hour late when I reached the church and my friend was very relieved when he saw me.

During my stay in Princeton I was visited once by a group of Greek friends. Among them there was a young lady, Miss Vaya Mantaka, who was later to become my wife.

I stayed in Princeton only for one month. After that I went, for two months, to the Institute of Space Studies in New York (1963).

There I had the benefit of some really good computers (Fig. 7). Among other things I calculated (with Bozis) orbits during the collision of two galaxies. I wanted to find the bridges that join two interacting galaxies. But I used a very high velocity of collision from the literature, and the results (Contopoulos and Bozis 1964) were not so spectacular. When I realized that, it was too late. I had left New York, and I had no more access to a large computer.

Figure 7. A happy person at the Institute for Space Studies with a lot of computer output.

A much more detailed study was made later by Toomre, who found some very nice bridges and tails in interacting galaxies. Toomre sent me a preprint of his paper where he wrote: "Contopoulos and Bozis (1964) reaffirmed that very hyperbolic encounters likewise build no bridges; it is amusing that their last remark was that perhaps much slower collisions may eventually explain the formation of bridges". I replied to him: "We went a-hunting with only one bullet and our single shot missed the game... Thus, for us, the fact that our results were negative and we could not continue our exploration of other cases, was sad and not amusing". Then Toomre wrote me: "I should have thanked you at once for your "a-hunting with only one bullet" letter of June 22 in which you objected, ever so politely to our use of the word "amusing". Then he changed "amusing" into the less ambiguous "ironic" in his published paper (Toomre A. and J. 1972).

At the Institute of Space Studies I did a lot of work on resonances in galactic dynamics. The director of the Institute was Dr. R. Jastrow and he was very generous to me. He had two secretaries and a draftsman at my disposal and I gave them quite some work to do. Furthermore, he invited a young post-doc to work with me. This post-doc had done his thesis with Chandrasekhar on oscillations of incompressible stars. His work indicated the existence of a third integral in this case. We discussed what should be done to promote this subject further, but then I had to leave for Greece. When I came back one year later I asked him what he had done during the past year. "Very little" he replied, "I did not know what to do further, so I was waiting for you to come back". I am sorry that this subject is left unexplored until today.

During my stay at the Institute for Space Studies I made arrangements to continue part of my calculations from Greece. A friend at the Institute kept my programs, in the form of decks of cards, and promised to do the calculations with data that I should send him from Greece. It is remarkable that when this friend had to leave New York he asked a colleague to continue helping me, and when this colleague also left, he asked a third person to help me. That method worked for several years (with occasional visits to New York). But finally the connection was lost. The last time I came to New York I had great difficulty in finding my decks of cards. Finally I did find about half of them in some remote drawers. But from then on, cards became obsolete and we used magnetic tapes. I always carried such tapes with me in later years.

In New York I had many Greek friends that I met mainly at the Greek orthodox cathedral every Sunday. The churches are the meeting places of Greeks all over the world. I organized several excursions with my friends, in the Catskills, the Adirondacks, the Long Island, etc.

One friend was Dr. Angelo Skalafuris who was working next to my office at the Institute of Space Studies. He helped me several times with his car. On one occasion I asked him if he could take me from the airport when I should return from a trip next Sunday. "At what time are you coming"? he asked. "About 5 o'clock" I said. He looked embarrassed. "I am sorry but I cannot come" he said "because at that time I am getting married". I offered him my best wishes and I added: "How is it possible to say that you are sorry, if you are getting married"? The coming Sunday I arrived on time and at 7 p.m. I was in my office at the Institute working. Some time later I heard someone typing in the next office. I could not believe it. I went to the next office and I saw

Angelo typing. I was greatly surprised. "Didn't you tell me that you would be married today"? I said. "Oh, yes", he answered "I was married o.k. But I had to finish a paper and so I came to the office". That was unbelievable. In all Greek families a marriage is an event of the greatest importance and it is celebrated accordingly. I remember a discussion I had once with an American colleague. He was preparing the marriage of his daughter and he was explaining to me how difficult it was to organize the wedding party. And then he concluded: "But after this party my obligations will be over". I replied to him: "You are lucky. In our case we have to provide at least a house for our daughter".

As I left New York, I flew directly not to Athens but to Ankara, Turkey! I was invited to a NATO summer school there, together with Lynden-Bell, Longair, and several others. The organizer of the school was Dr. E. A. Kreiken, a Dutch professor at the University of Ankara. His association with the University of Ankara was rather strange. In fact Dr. Kreiken was giving his courses in German and a student translated into Turkish!

Prof. Kreiken wanted to have the lectures of the summer school printed as a book. I had only one copy of my lectures (unfortunately I did not keep a carbon copy and there were no xerox machines at that time). With some hesitation I gave my copy to Dr. Kreiken and he promised to send me the proofs soon. Unfortunately a short time later Kreiken passed away. I wrote several letters to ask about the fate of my manuscript, but never received a reply. I was convinced that my manuscript was lost. Then 20 years later I met Russell Cannon from Edinburgh, and we remembered the school in Ankara. I told him my complaints about my manuscript and Russell exclaimed: "But the Proceedings did appear". Some time later he sent me a photocopy of the book, that had been published by the University of Ankara in 1974, 11 years after the school, without ever notifying me!

10. FIRST IAU SYMPOSIUM IN DYNAMICAL ASTRONOMY

When I returned to Greece, after one and a half year in the U.S.A, I was engaged to Miss Vaya Mantaka, who had meanwhile returned from the United States to Greece. I met her mother (her father had been executed by the Germans during the Occupation, because he was involved in the underground resistance) and we went together to my family in my native town of Aigion, where my father passed the rings.

The year 1963-1964 I was fully occupied with teaching and writing of several papers. At the same time we were organizing the first meeting in dynamical astronomy, namely the IAU Symposium No 25 on the "Theory of Orbits in the Solar System and in Stellar Systems" (Figs. 8, 9).

The chairman of the Scientific Organizing Committee was Prof. D. Brouwer. There were 65 participants from 13 countries, and I was the Editor of the Pro-

Figure 8. The participants of the IAU Symposium 25 on "The Theory of Orbits" as they arrive in Thessaloniki (1964).

Figure 9. My fiancée with Y. Kozai, Y. Hagihara and G. Hori in Thessaloniki (1964).

ceedings. Among the participants there were, besides the Greeks, S. Aaserth, G. Clemence, J. M. A. Danby, M. Davis, A. Deprit, W. Eckert, B. Garfinkel, P. Goldreich, Y. Hagihara, M. Hénon, S. von Hoerner, G. Hori, I. King, J. Kowalevsky, Y. Kozai, M. Lecar, D. Lynden-Bell, R. Miller, A. Ollongren, L. Perek, K. Prendergast, M. Schmidt, V. Szebehely, S.M. Ulam, J. Vinti, and R. Woolley. Ulam in particular (the one of the Fermi-Pasta-Ulam problem) was not an astronomer. He came all the way from the United States for the Thessaloniki Symposium, and then he returned home. He did not go afterwards to the IAU General Assembly in Hamburg, like all the others.

The meeting was a real success. We had brought together people working in celestial mechanics and in stellar dynamics. The two groups of people looked initially like two non-interacting fluids. Each group spoke their own language and terminology, and they had difficulty in understanding each other. But after this meeting the interaction increased considerably. Stellar dynamicists started using perturbation methods, as in celestial mechanics, and celestial mechanicians started using statistical methods, as in stellar dynamics.

During this Symposium we had several papers on orbits and integrals of motion in galaxies, on the N-body problem and the dynamics of clusters, on the three-body problem, and on the motions of planets and satellites, including artificial satellites. Some of the papers opened really new ground in several fields.

The meeting was successful also from the organizational point of view. Most participants stayed in the students' hostels paying 1 dollar a day. When the others, who had reserved rooms in nearby hotels, saw the excellent conditions in the students' hostels they left their hotels and moved there.

Every day after the morning session we had busses that took us to the beach. We could swim and then have lunch in a nearby restaurant. The only difficulty was to bring the participants back for the afternoon session. I had to swim far away to inform people that the busses were leaving soon.

In the evenings we had a few exceptional events, like a boat ride in the moonlight with music from guitars. We had also a number of excursions, around Thessaloniki (Fig. 10) and later around Athens. During the closing dinner in Thessaloniki Victor Szebehely gave a speech. He said that the meeting emphasized Contopoulos' third integral, and he added: "For the benefit of the ladies I must explain what is the third integral. It is something very important that has two remarkable properties. First it is not an integral and second it is not the third". It was crystal clear!

I prepared the proceedings very fast after the meeting (Contopoulos 1966c), but I lost an important paper, the one by Prendergast, on barred galaxies. He gave me a manuscript in very bad shape, with notes written by pencil, and sketches instead of figures. I asked him: "Can you provide a better manuscript soon?" "Yes", he said. I gave him back the manuscript, expecting a better one.

Figure 10. With M. Hénon at an excursion in Pella during the IAU Symposium No 25 (1964).

That was my mistake. I never saw the manuscript again, nor did any one else. Prendergast was a genious but also known for his neglect in publishing papers. Unfortunately I did not know it at that time.

I was unfortunate also with another paper, presented by a well known astronomer, that dealt with a multivalued potential (!) (proportional to exp (2θ)). Someone asked what is the density and the author replied that it is found by Poisson's equation. I pointed out to him that this gives a multivalued density, but the author replied that he could not offer a better answer.

One year later Dr. Brouwer organized another meeting in Stanford, California (Fig. 11). It was also a very good meeting. But when we discussed it in front of Mrs. Brouwer, she interrupted us and said: "There is no comparison with the Contopoulos meeting. There everyone was taking care of the ladies!".

Many meetings of the same type followed in later years. But the Thessaloniki meeting was always remembered as the first meeting on dynamical astronomy.

11. IAU COMMISSION 33

After the IAU Symposium No 25 we had the IAU General Assembly in Hamburg (1964).

There I had an unexpected and happy surprise. Bart Bok, who was then president of Commission 33 (Structure and Dynamics of the Galaxy), a successor of Oort and Blaauw, proposed me as vice-president of the Commission, implying that I should be his successor as president after three years. But his suggestion was objected by the Executive Committee. Bok told me that I had a serious drawback. I was not Dutch. "But, I told him, you, who proposed me, are Dutch". "Not the usual type of Dutch", said Bok and he added "I know what

Figure 11. A group of celestial mechanicians at Stanford, California (1965). From the left: P. Message, A. Deprit, B. Garfinkel, T. Danly, D. Brouwer, M. Davis.

I will do". He secured first a unanimous vote of the Organizing Committee of the Commission in favour of Contopoulos, and then he went to the Executive Committee and said: "The Commission will not accept another vice-president except Contopoulos". So the Executive Committee accepted the verdict and I was appointed vice-president. Later, at the Prague General Assembly of the IAU, I was elected president of Commission 33 without any problem.

The main innovation that Bok and I introduced in Commission 33 was a "long report" of the Commission, that was published independently of the short report, the one that was included in the IAU Proceedings. This short report, for reasons of lack of space, had an almost telegraphic style, with a long list of references. The longer report gave us the possibility to include an evaluation of the various contributions to our field, and more extended references. Thus, this report was greatly appreciated everywhere.

The tradition of the long report, initiated by Bok, continued for many years. When I was president of the Commission, the long report was published as a book by the University of Thessaloniki. Some people called it "a very nice and extremely useful Report" (P.O. Lindblad), "the finest IAU Commission Report I have ever seen" (W.W. Morgan), etc.

At the General Assembly there were several young Greek astronomers. During a boat trip along the river Elbe we gathered together on the deck and we organized a choir (I had some assistants that were excellent singers). We sung several Greek and classical songs. While we were singing many participants came to the deck and were listening, seemingly looking at the water. When we finished we were surprised to hear a general clapping of hands. Our songs were melodious, but rather sad, especially the early popular Greek songs, that originated before the Greek revolution. K. Prendergast was saying: "There is nothing more melancholy, than three happy Greeks singing".

Whenever I was going to a meeting abroad, I tried to give myself the luxury of one day of sightseeing, after the meeting was over. What I liked most was mountaineering. Nothing sophisticated, of course, I just walked in my regular shoes up the mountains, as high as I could. I liked in particular the Austrian and Swiss Alps. I remember one occasion, when I had climbed a peak near Matterhorn. I was alone. Then a group of tourists arrived, dressed in proper alpinist style, and they asked me: "To what alpinist club do you belong?" I had no idea about alpinist clubs at that time.

After the Hamburg General Assembly I had my marriage in Athens. Then we went on a honeymoon in Germany (Munich and Garmisch) and Austria (Innsbruck). We arrived in Munich during the Octoberfest holiday. We went to a large beer hall dressed in Bavarian costumes, trying to be inconspicuous. But the people sitting next to us greeted us smiling and said: "Welcome. When did you get married?" We could not avoid being conspicuous as foreigners and newlywed. Then we went mountaineering near Innsburck and we had a very nice time in the Austrian Alps.

12. APPLICATIONS OF THE THIRD INTEGRAL

For many years after 1960 I worked on the third integral (section 5). The third integral has a large number of important applications. One of the first was to find the boundaries of the orbits (Contopoulos 1960). The box orbits fill curvilinear parallelograms, while the tube orbits fill tubes around particular periodic orbits (Fig. 4). In both cases the boundaries can be found using the appropriate form of the third integral (Contopoulos 1965a). A particular application was to find the periodic orbits and their stability character (stability, or instability). This method gives very good results, that can be compared with the empirical calculations of periodic orbits, by using a Newton method, or any other numerical method (Contopoulos 1965a, 1968). Some other applications of the third integral are the following:

12.1 Celestial Mechanics

I developed the third integral in the restricted three-body problem, the elliptic restricted three-body problem, and the 3-D restricted problem (Contopoulos 1965b, 1967a,b). These integrals are important in explaining the forms of the orbits found numerically (see e.g. Bozis 1966, 1967). Very recently the integrals of the elliptic restricted three-body problem were used in explaining the appearance of stable chaos in the solar system (Varvoglis et al. 2003). Namely the orbits of certain asteroids are chaotic, but nevertheless these asteroids do not escape from a certain resonance region. The explanation is that these orbits are restricted by constraints due to a third integral of motion.

When the use of computer algebra was generalized it was possible to find series expansions for the third integral to very high orders. A particular application refers to the orbits of the Trojan asteroids near the Lagrangian points L_4 and L_5 in the solar system. These points are the apices of equilateral triangles with the other two apices at the Sun and Jupiter. Near these points there are many asteroids.

The Lagrangian points L_4, L_5 in the Sun-Jupiter system are stable. Near these points a third integral secures the stability of the orbits in this neighbourhood. What was not certain is the extent of this neighbourhood. An application of the original KAM theorem (of Kolmogorov, Arnold and Moser) by Hénon (1966) gives stability only up to a distance of the order of one millimeter. An extension of the KAM theorem is Nekhoroshev's theorem that finds stability for a certain long time (the age of the universe) and not necessarily for all times. Using this theorem and calculating higher order terms of the third integral Giorgilli et al. (1989) increased the stability region up to a few kilometers. Only in recent years, by calculating terms of the third integral up to order 30 it has been possible to extend the stability region to about 0.3 Astronomical Units (Skokos, Contopoulos and Giorgilli 1996, Giorgilli and Skokos 1997). In this region we find, for the first time, real astronomical objects, i.e. a few real Trojan asteroids. More recent work (Efthymiopoulos 2004) extended the domain of applicability to include several tens of asteroids.

Therefore the third integral and the theorems about its applicability (KAM and Nekhoroshev theorems) can be applied to real astronomical objects, if terms of sufficiently high order are calculated.

12.2 Galactic Dynamics

The first application of the third integral was in explaining the triaxial form of the velocity ellipsoid of the stars near the sun. It is known that the distribution of the velocities of the stars in the solar neighbourhood follows an ellipsoidal law, i.e. the equidensity surfaces are concentric similar ellipsoids. The radial axis of the ellipsoid is the longest, while the z-axis (perpendicular to the plane of symmetry) is the shortest. However, if there are only two integrals of motion, namely the energy and the z-component of the angular momentum, these two axes should be equal. Only the introduction of a third integral allows the two axes to be different (Barbanis 1962).

Further applications of the third integral refer to the types of orbits in a galaxy. In recent years there have been many N-body simulations, representing the collapse of a protogalaxy, that leads to the formation of an elliptical galaxy. The final form of the galaxy is a self-consistent model, that remains approximately stationary. The distribution of velocities then remains constant. The use of the third integral in such a model not only describes the forms of the orbits, but gives also their velocity distribution and explains how the self-consistency of the model is achieved (Contopoulos, Efthymiopoulos and Voglis 2000; Contopoulos, Voglis and Kalapotharakos 2001).

A particularly important application of the third integral in galactic dynamics is the nonlinear theory of density waves. In section 15 we will see some developments of this theory between 1970 and 1980.

12.3 The Fermi-Pasta-Ulam Problem

This problem deals with the dynamics of N coupled oscillators. It was studied numerically by Fermi, Pasta and Ulam (1955) by using the very early computers of that time. The coupling term was taken to be small. The authors expected that, as the coupling term made the system nonlinear, the energy would be shared equally by all oscillators. Instead of that, when the energy was given initially to one oscillator, only a small part of it was later shared by a few oscillators, and this was done in a periodic way. All the energy returned to the original oscillator after a certain period of time, and the process was repeated.

That was a paradox, but it can be explained if we notice that the energy of each oscillator is the first term of an integral of motion of the third integral type. The higher order terms represent the variations of this energy, but, when the variables come back close to their initial values, this energy recovers its original value. This explanation of the Fermi-Pasta-Ulam paradox (Contopoulos 1966c) has been proposed also by many other authors.

However, it was noted that if the coupling is large, or if the energy is large, the energy sharing between oscillators becomes almost complete. The main problem is then to find the critical coupling (or critical energy) beyond which the oscillators exchange freely their energy. This subject is related to the "resonance overlap" problem that is discussed in the next section.

12.4 The Störmer Problem

The trapping of charged particles in the magnetic field of the Earth is known as the Störmer problem. Such particles stay in the well known van Allen belts, far above the Earth's atmosphere. But sometimes these particles come to lower altitudes near the poles of the Earth, producing the phenomenon of aurorae.

A theoretical study of the motions of charged particles in the dipole magnetic field of the Earth has been made in particular by Dragt and Finn (1976, 1979).

This subject was the thesis of J.M. Finn. I followed with much interest Finn's work while I was a visiting professor at the University of Maryland.

I studied myself a particular case of the Störmer problem, namely the case of resonances (Contopoulos and Vlahos 1975). In all cases a third integral plays a basic role in the orbits of particles. But the third integral is different for each resonance. Therefore one has to be careful in applying the third integral in every case.

12.5 Solitons

Certain partial differential equations (PDEs) have some remarkable solutions that represent isolated waves called solitons. Such equations are the Korteweg-de Vries (KdV) equation, the sine-Gordon equation, the nonlinear Schrödinger equation, etc. These are called "integrable PDEs".

The solitons move indefinitely unchanged, until they are reflected. When two solitons collide they interact in a very complicated way, but later on they reappear unchanged as they move away from each other. Solitons appear in water waves, in plasma physics and in many other physical systems.

The theory of solitons of the Korteweg-de Vries equation, was developed around 1965 by Martin Kruskal, Norman Zabusky and others in Princeton (e.g. Zabusky and Kruskal 1965, Miura, Gardner and Kruskal 1968; see also the review by Scott et al. 1973). They found an infinity of conserved quantities that behave like integrals of motion.

Dr. Kruskal was giving a seminar on this subject at MIT in 1969. During the discussion period I pointed out to him that his conserved quantities looked very much like the third integral. He said that he could not see a clear connection. But a few years later it was proven (Zakharov and Fadeev 1971) that the Korteweg-de Vries equation represents a completely integrable system of infinite degrees of freedom, with infinite integrals like the third integral. In fact the KdV equation is the limit of the Fermi-Pasta-Ulam lattice when the number N of bodies tends to infinity (Cercignani 1977).

What is most interesting is that perturbed "integrable PDEs" have solutions that are soliton-like, although not exactly solitons. These are isolated waves that change slowly in time, but nevertheless they behave like solitons over long times. In recent years a more direct relation between the third integral and particular solitons was established in galactic dynamics (Voglis 2003).

12.6 Other Applications

Dragt et al. (1988) developed a theory equivalent to the third integral to describe the motions of charged particles in an accelerator. In this case one has to consider the effect of many magnets on the motion of the particles and this can be done only with computer algebra. Such work is absolutely necessary in studying and programming large accelerators.

Lovelace (1978), and several other people, have applied a theory of integrals of motion to various machines used for plasma confinement (astron, tokamak, etc). The literature on this subject is so extensive that we cannot give even representative references here. This subject has important applications in controlled fusion.

Similar phenomena apply to molecular dynamics. A representative paper by Uzer et al. (1991) has the title "Celestial mechanics on a microscopic scale". Of special interest are phenomena related to atomic and molecular physics. The relation of this work with quantum mechanics will be discussed in section 37.

Other applications refer to nonlinear electric systems, to phenomena near the absolute zero of temperature, and to the dynamics of elementary particles.

As regards astronomy, one should mention the application of integrals of motion to the pulsations of variable stars (Perdang and Blacher 1982, 1984) and to the motions of galaxies in cosmology. The applications of the third integral in relativity and cosmology will be discussed in section 36.

12.7 The Breakdown of the Third Integral

When the perturbations of an integrable system become large the third integral breaks down. That means that the orbits become chaotic and are not constrained by any integral besides the classical integrals of energy, angular momentum, etc.

The transition from a system that is mostly ordered to a mostly chaotic system is quite abrupt. If the perturbation goes a little beyond a critical value, the system becomes very chaotic (Hénon and Heiles 1964).

This transition to chaos is due to a phenomenon called "resonance overlap". I.e. if the perurbation is sufficiently large the orbits are influenced by more than one resonance at the same time and they become chaotic. This phenomenon is described in the next section.

13. RESONANCE OVERLAP

During the summers of 1965 and 1966 I was a research associate at the Institute for Space Studies in New York. On both occasions I made several trips all over the United States, and gave many lectures.

In 1965 I attended a summer school on dynamical astronomy in Stanford, California. Then I organized, with Prof. Strömgren, a workshop on galactic models, in New York. I attended also a summer seminar on relativity theory and astrophysics in Boston, where I had my first contacts with the work of C. C. Lin and his associates on spiral density waves.

In 1966 I attended a conference in Liège, Belgium. There I presented my first calculations of orbits in a realistic spiral field (Contopoulos 1967c).

After the Liège conference I travelled to New York. During this trip my luggage was misplaced and I did not find it when I arrived at the New York airport. When I reached the Institute for Space Studies I found a great excitement. Most of the personnel would travel this night to Cape Kennedy to see the launching of a satellite next day and they invited me to go with them. I was glad to accept, although I did not have even the most necessary items from my luggage. We reached the Cape next morning and we came very close to the launching site. It was a very impressive sight, especially during the countdown. Unfortunately the launching was postponed the very last second. Then we had a long journey north by train. And I was fortunate enough to find my luggage on my return, waiting for me in my office at the Institute for Space Studies.

Later this year I organized a workshop on spiral structure in New York. The participants were B. Strömgren, C.C. Lin, F. Shu, A. Kalnajs, A. Toomre, P. Vandervoort, K. Prendergast, M. Lecar and myself. Frank Shu was taking notes. These notes were never published, but I still have a copy of them.

The discussions during this meeting were the precursors of the later developments in this field. In particular we had the first estimate of the spiral pattern velocity in our Galaxy, given by Strömgren, who found the positions of the spiral arms where stars of our neighbourhood were born. From these past positions and the present positions of the spiral arms he derived a pattern velocity of 20km/sec/kpc, which is of the right order of magnitude.

Another meeting that I organized this year in New York was devoted to the N-body problem. At this meeting Dr. M. Lecar proposed that all people doing N-body experiments should do the same calculations (the same initial conditions of 25 bodies) in order to compare the results (section 16).

During the summer of 1966 we had a symposium in Besançon on "New methods of stellar dynamics". Hénon and Nahon were the editors of the Proceedings, which were published by the Centre National de la Recherche Scientifique (CNRS) of France. I flew to France from New York for this meeting and returned to New York afterwards. I presented my recent study of the interaction of resonances, which produces chaos in a galactic system (Contopoulos 1966d).

I had already developed the theory of resonances (1963), which explains the various forms of the tube orbits (see section 7). Every resonance requires a different form of the third integral. But when the perturbation increases, the resonant regions increase in size, and they finally interact with each other. E.g. in a 4:1 resonance one finds four islands of stability, and in a 3:1 resonance three islands of stability (Fig. 12a). The sizes of the islands are derived theoretically by the corresponding resonant forms of the third integral. But if the perturbation is large enough the theoretical islands should overlap. In fact the orbits cannot follow one resonance rather than the other and they become chaotic. Then the invariant surfaces that separate the various resonances are destroyed (Fig. 12b). The interaction of resonances is the basic mechanism of generation of chaos. One resonance alone does not produce chaos. In fact there are resonant integrable dynamical systems with no chaos at all. But, when the orbits are influenced by more than one resonance, they behave irregularly, in a chaotic way. During the year 1965-66 I calculated, theoretically and numerically, the limits of the resonant regions in a simple dynamical system, representing two coupled oscillators. I could find analytically the critical perturbation required for an interaction of resonances (resonance overlap) and that agreed well with the numerical calculations. My results were presented for the first time (1966) at the Besançon Symposium (Contopoulos 1966d). The Symposium proceedings were published later (1967) in the French journal "Bulletin Astronomique".

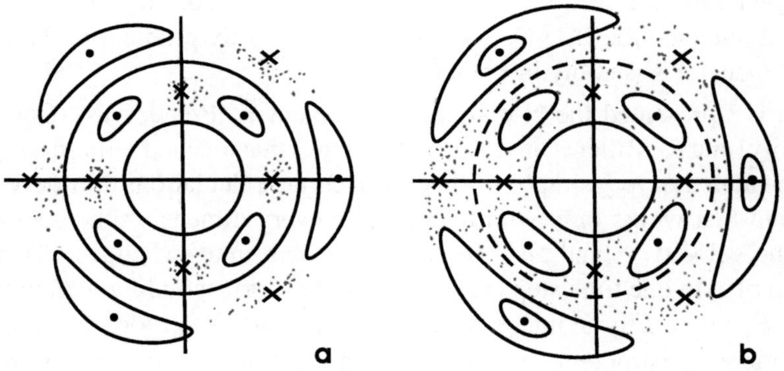

Figure 12. Resonance overlap. When the perturbation increases the small chaotic domains of multiplicities 3 and 4 of (a) join and form a large chaotic domain. (b)

I realized later that a group of people in the USA (Rosenbluth et al. 1966) made a similar estimate of the resonance overlap the same year (1966), by working on the "destruction of magnetic surfaces by magnetic field irregularities".

A detailed discussion of the resonance overlap was made in 1979 by Chirikov in Physics Reports (1979). His paper is quite clear, and became well known. Because of that many people call the resonance overlap "Chirikov's criterion". But Chirikov himself in an earlier paper (Zaslavsky and Chirikov 1972) gives credit to the previous papers of Rosenbluth et al. and of Contopoulos, with the following words "The criterion of stochasticity (2.11) implies overlap of resonances of the same order of magnitude. In the general case, stochastic instabilities can arise from intersection of narrow stochastic layers... The criterion of such weak stochasticity has been studied in [64,65]" (64=Rosenbluth et al., 65=Contopoulos).

Some people refer to a short earlier paper of Chirikov (1959). But this paper considers the successive crossings of various separate resonances by orbits, without mentioning resonance overlap. Later Chirikov did consider resonance overlap and in a letter to me, dated 23-10-68, he wrote: "My criterion of stochasticity by resonance overlapping is essentially the same as yours in your paper in Bulletin Astronomique (1967)".

At the Besançon Symposium there were many lively discussions, that demonstrated the vitality of the "new methods of stellar dynamics". Among other topics there were calculations demonstrating the Lynden-Bell statistics, several N-body calculations, etc.

14. THE JUNTA PERIOD AND THE IAU

The year 1967 was also very busy. Besides the New York Institute for Space Studies, I attended the IAU Colloquium on the N-body problem in Paris, and the IAU General Assembly in Prague.

In Paris I presented the N-body calculations of K. Prendergast (Columbia University) and R. Miller (University of Chicago) that reached millions of particles. The idea was very simple. Instead of accurate calculations they developed a program that works only with integers. In every dimension the coordinate x and the velocity v (and also the time) take only integer values. Then the variation of x in one time step is $\Delta x =$v, and the corresponding variation of v is Δv$= F$, where F is an integer force (acceleration). Therefore the motion of large numbers of particles does not involve any calculations, but only transfer in the memory of the computer, by integer values of v, or F. The only calculations are made in estimating the force F. This allows a spectacular increase of the capability of computers in dealing with the N-body problem.

In Prague I organized a meeting of Commission 33 on the "theoretical approaches to spiral structure". This contained papers by C. C. Lin on "the density wave theory", and by S. B. Pikelner on "magnetic approaches to spiral structure". I presented also a gas dynamical version of the density wave theory that was developed by M. Fujimoto in Columbia University. I was familiar with this work since I had been in Columbia several times and had many discussions with Fujimoto and Prendergast. (The work of M. Fujimoto was published later (Fujimoto 1968)).

Various attempts to explain the spiral arms as due to magnetic fields in galaxies had been made over the previous years. But all such attempts failed. One reason was that the magnetic field of our Galaxy is too weak to have any appreciable large scale effects. The other reason is that no credible theory could explain the spiral arms as due to magnetic effects. On the other hand the density wave theory was extremely successful in explaining the spiral arms as density waves rather than material arms. In the program of the Prague meeting I included the paper on the magnetic theory of Pikelner, in order to be as impartial as possible. But from then on all magnetic theories vanished into oblivion.

At the Prague General Assembly of the IAU we heard an invited lecture by M. Ryle (who won later (1974) the Nobel prize) on Cosmology. He emphasized the main arguments in favour of the Big Bang. His lecture was very nice, but at that time the opponents of the Big Bang were still strong. I remember that G. Burbidge, a supporter of the steady state theory, spoke one day after M. Ryle and said: "When I heard Dr. Ryle I felt like... being in a church!" (implying that the Big Bang theory was influenced by religion).

But a few years later Y. Zeldovich, in his invited lecture at the Patras General Assembly (1982), stated that "the hot Big Bang theory has been established today beyond any reasonable doubt" and he compared this theory with the "realization that the Earth and other planets move around the sun" (Zeldovich 1983).

The year 1967 was crucial for me from another point of view. A dictatorship was established in Greece, and many of my activities became very difficult. Whenever I had to travel abroad I had to get permission from the police authorities, but even when I entered my plane I was not sure that I should be allowed to travel (a colleague of mine was taken out of his plane the very last moment). Even when one was abroad he was not quite safe if he wanted to return to Greece. I remember that at a meeting in Boston a gentleman approached me and introduced himself. "I am the Greek consul in Boston" he said. "You did not come to visit us". I replied that I was very busy. "But your work did not prevent you from visiting Mr. so and so, etc". And he gave me an accurate list of all my Greek connections. They knew exactly what I was doing in Boston.

On another occasion I had taken an American friend to a Greek restaurant in New York. During the dinner I was telling him what we were suffering in Greece because of the Junta. Then at one moment I heard my name "Mr. Contopoulos!" I turned around and I saw a colleague of mine from the University of Thessaloniki, who was sitting exactly behind my chair. This person was well known for his support of the Junta, and was appointed later minister of education by the Junta. He had obviously heard every word I was saying.

Fortunately, I was not persecuted seriously by the Junta. But other colleagues of mine were arrested, or exiled, and on one occasion many people lost their jobs. When this happened I proposed to my colleagues to give our resignations to our Rector, so that he should try to convince to ministry to retract its decision to fire our colleagues. Some of my colleagues followed my example and the ministry did promise to reconsider its decision. Somehow the foreign press learned about our action, and that was made widely known after a broadcasting from the BBC. After that the Dean of our faculty told me that the chief of the secret police visited him and told him to warn Contopoulos to be careful. He said: "Contopoulos says that he does not care if we fire him because he will find a position abroad. But tell him that we will take away his passport and he will not even be allowed to travel abroad".

From that time on I had many harassments and threats. On one occasion I received an unsigned letter, from someone who was threatening to kill me, unless I passed all the students in the coming examination. In the previous examination I found that someone had cheated. I told my assistants to call him to my office. An assistant knew him and warned me: "Sir, this person is a high ranking officer of the secret police". But I insisted. He came to see me in an official uniform with all his decorations. I told him that he was failing this exam, and he did not dare to say anything.

The day before receiving the threatening letter I had a telephone call from the police. "Did you receive a letter threatening to kill you?" I was surprised "No" I said. "You will receive it. But do not be afraid. We will be guarding you in a discrete way". I asked: "Who is calling, please"? He did not want to give a name. I said: "How do I know that it is not a farce?" Then he gave me a telephone number. The next day I did receive the threatening letter and I called the police. I feigned to be very afraid. I asked: "How did you know about that letter?" Finally they told me: "You know, we are ordered to open the letters of some people, including you. But do not be afraid". I was afraid not because of the threatening student, but because I had written a number of letters abroad denouncing the practices of the Junta. In particular, I had written a long letter to Chandrasekhar. As I never received any reply I wondered whether my letter was kept by the police. I did not dare to write another letter. One year later I met Chandra and I asked him if he had received my letter. "Yes", he said, "and I felt so sorry, that I could not write anything in reply"!

Then I had an unexpected support from abroad. I received a letter from Dr. O. Heckmann, President of the IAU, asking me to visit him in Hamburg during my next trip abroad. He invited me to give a lecture in Hamburg and he covered all my expenses. He wanted to ask me if I would accept to be the next Assistant General Secretary and, later, General Secretary of the IAU. I was happy to accept, but I warned him of the political situation in my country. He thought that if I had such a position the Junta would not dare to harm me.

My official election as Assistant General Secretary took place at the IAU General Assembly in Brighton (1970) (Fig. 13).

My first initiative in Brighton was to propose the formation of a European Astronomical Society. I had strong support from the representatives of the main European countries (United Kingdom, France, Germany and the Soviet Union). But when the Soviet representatives returned to their country, they met an absolute refusal from their authorities. They sent me a friendly, but negative

Figure 13. The old and the new IAU Executive Committee at Brighton: M. Schwarzschild, L. Gratton, L. Perek, W.N. Chrisrtiansen, P. Swings, J. Sahade, O. Heckmann, Sir B. Lovell, B. Strömgren, B.J. Bok, C. de Jager, A.B. Severny, G. Contopoulos, M.K.V. Bappu, J-C. Pecker.

Figure 14. Renewal of the IAU Executive Commitee at Grenoble (1976): C. de Jager, B. Strömgren, G. Contopoulos, W. Ivanoswska, J.G. Bolton, L. Goldberg, E.K. Kharadze, C. Fehrenbach, E. Müller, P.O. Lindblad, B. Bok, Sir B. Lovell, A. Blaauw.

letter. Thus this project stopped at that time and only many years later (1990) a European Astronomical Society was finally formed[5].

[5]A group of former General Secretaries of the IAU, including myself, elected the first council of EAS and L. Woltjer became its first president (1990). The society has now some thousands of members and it is very active.

Instead of a European Astronomical Society I made then another proposal. To have Regional Meetings of the IAU, in Europe and in other parts of the Globe. This idea was highly successful and we had many regional meetings, starting with a meeting in Athens in 1972.

In Brighton I faced a serious problem. The Executive Committee of the IAU decided to move its secretariat in Paris. Until then the secretariat moved to the institute of each General Secretary. But this was considered impractical. Therefore, they wanted to establish a permanent secretariat in Paris during my term. I considered that impossible for me because I would have to establish a Greek secretariat to communicate with my secretariat in Paris, and I thought that this would be impossible, especially under the Junta regime. Thus, immediately after my election I said that if this was the decision of the Executive Committee I had to resign from my post. Finally, the Executive Committee agreed to postpone the establishment of the Paris secretariat, and the problem was solved at that time.

Before taking over the position of the Assistant General Secretary I visited the minister of education and the rector of our University. Both said that they would support me in all possible ways. But they would not provide any written commitment. Then the minister changed. And a new rector was appointed by the Junta (who was not elected by the professors, as in the past). Nevertheless I could continue my work, but with many difficulties. To give a few examples: I had to submit a list of all the letters that I mailed to get refunded. I had two secretaries, but they had no experience in formulating even the most simple letters. Thus, I had to write all the letters myself and then check carefully the typed version. There were much better secretaries outside the University, but they were paid higher salaries, that I could not afford to pay.

The situation improved considerably when I became General Secretary. Then I had four times the previous work, but the IAU provided two very experienced secretaries, Mr. A. Jappel and Mrs. J. Dankova, that took care of most of the correspondence.

At that time I even succeeded in getting some financial support from the ministry for the meetings of the IAU officers and of the Executive Committee. But I had again problems when the minister changed. In one case I visited the new minister and told him that I had the promise of support from the previous minister. His reply was that he was not bound by any promises of the previous minister. I went out very unhappy. Then I met in the corridor the general secretary of the ministry, whom I knew already, and he asked me: “Why are you so sad”? I told him and he said: “Don’t worry. I will take care of that matter myself”. The next day he called me and told me that my request was approved. “How?” I said. “I just wrote a positive reply and the minister signed it without asking questions”.

15. SECOND SABBATICAL IN THE USA. THE DENSITY WAVE THEORY

During the Junta period I considered the possibility of emigrating to the United States. I had invitations to go as visiting professor at the Universities Columbia, Harvard and MIT, and an offer of a full professorship from the University of Chicago. I asked for a sabbatical leave of absence, and left the country with my wife and two children (May 1969). When I arrived in Columbia, I received a telegram that all leaves of absence were cancelled and I had to return immediately. I replied that I could not return. Then, I learned later, there was embarrassment in the ministry. They considered seriously of firing me, but finally they decided not to do so.

In Columbia I could do very good scientific work. I had many useful scientific discussions with Lo Woltjer (director of the astronomy department), and Kevin Prendergast.

When Lo Woltjer was, later, Director General of ESO, he invited me several times there as a research associate. A characteristic of his character is the following. When I wanted to see him I called his secretary to ask if I could visit him. In general he was busy at that time. But when he was free, he would not invite me to his office (as every other Director was doing) but he would come to see me in my office.

In Columbia he was Editor of the Astronomical Journal. He improved the standards of the Journal considerably, being quite strict with the refereeing system.

On the other hand, Kevin Prendergast was a hidden treasure. It was surprising how many things he knew in mathematics, physics and astronomy. He would often give a crucial advice on very different subjects. His only (but serious) drawback was his neglect in publishing his papers. Besides the paper on barred galaxies that was supposed to be published in the Proceedings of the Thessaloniki Symposium but never appeared (section 10), he had many more ideas that were really lost. He had developed an impressive statistical theory of the three-body problem, that was never published. Then he developed a theory of rational approximations (similar to the Padé approximations, but better) that could give the periodic orbits of a dynamical system with much better accuracy than various series expansions like the third integral. He published a paper on this subject in a Proceedings volume (Prendergast 1982), and several extensions or applications of his work were published by his students, or friends (e.g. Contopoulos and Seimenis 1990). However, most of his own work is in the form of preprints, or just notes.

I was surprised to see a number of papers by K. Prendergast, co-authored by G. and B. Burbidge, but these papers were due to the care of the Burbidges. They told me that in one occasion the Burbidges had a reference to a forthcoming paper by Kevin, in the form (Prendergast, in press 2010, or some other equally remote date), that was unfortunately eliminated by the Editor of the Journal.

Another characteristic of Kevin Prendergast was his neglect of secondary things. On one occasion I visited him in Columbia, coming directly to his office with my luggage after a long trip by plane. I expected him to say, first, the usual polite things about my trip etc. Instead, after we shook hands, he went directly to the blackboard and said: "I want to ask you a question about this formula...". We discussed this matter for hours and then I realized that it was well beyond lunch time and I was really hungry. I told him that, and he replied: "Oh! You want to have a lunch. O.k. Let's go somewhere" (that meant usually a Chinese restaurant).

In Columbia I stayed only one month. I was then working on dynamical systems with large (instead of small) perturbations (see section 22). I found numerically many interesting new phenomena, and I could explain some of them. But there are still some problems of highly perturbed systems that are open today.

When I was leaving Columbia I recommended to Drs Woltjer and Prendergast an assistant of mine, Dr. B. Barbanis, who was working with me on the third integral, and they invited him to Columbia.

When I returned to Greece I told Barbanis about the invitation. Barbanis was shy and his English was not perfect. He told me: "I am very grateful for your recommendation. But I will never forgive you for forcing me like that". Nevertheless his stay in Columbia was very successful. Among other things he published two interesting papers, one with Woltjer, and the other with Prendergast. And this visit proved to be a good asset for him when he was later elected professor.

After Columbia I went to Harvard, where I met my family (my wife and two children). My life there was very different from before. One of my first duties was to rent a house for my family. Then, I bought a car and my wife was driving because I had no driver's licence. Every evening she would park the car in front of our house, without looking for an appropriate parking place, as she was very tired. Then the next morning we invariably found a ticket on our car. I remember that we had a snowstorm in Boston, that covered all the cars deeply in the snow. Only the antenna of the radio was protruding. But next day I found the usual police ticket tied on this antenna.

After I paid several tickets I realized that I had to do something. I had never driven a car before. But moving carefully the various levers I could move the car slowly and park it safely. This is how I learned driving.

The car was a real blessing. We went around, to the sea and to the mountains, and the children enjoyed that very much. Only the beginning was difficult. We had bought our car in December 1968. Then some friends suggested to go to Vermont for Christmas. We would stay at the house of a friend there. There was snow everywhere. We went to the Greek orthodox church for the midnight service and it was wonderful. When we returned home we heard on television the greetings of the astronauts that were circling the Moon at that time. An astronaut read the first passages of the Bible: "In the beginning God created the Heaven and the Earth... And God said, Let there be light: and there was light. And God saw the light, that it was good..." and they finished by saying how happy they were to return now to the Earth. Then we had our Christmas meal and sung Christmas songs with the accompaniment of a harmonica.

After that we all went to the basement to sleep with many blankets. I was the last person to go to bed and I turned off the lights. Two hours later we felt a terrible cold to creep in. We got up and decided to wake up the lady-owner of the house, thinking that she had turned off the heating. She came, looked at the switches, and said: "Who has turned off the general switch?" Then I realized that I had turned off not only the lights, but the heating also.

In the morning it was sunny and beautiful. We went to a ski resort nearby, but the ski-lift was not operating. We asked why, and we learned that when the temperature was below -40°C the lifts had to stop. But it was only when I got out of the car that I realized how cold it was. We felt the cold as pain, and the children were crying.

During my stay at Harvard I was giving a regular course on galactic dynamics, replacing D. Layzer who was on leave of absence. Later on I was invited by C. C. Lin to MIT, and during my stay there I also gave a series of lectures. I had many discussions with C. C. Lin, Alar Toomre, Frank Shu and Agris Kalnajs on spiral density waves. My own work was on the nonlinear theory of density waves. The nonlinear theory is absolutely necessary near the main resonances, where the linear theory fails. Far from the main resonances of a galaxy (inner Lindblad, corotation, and outer Lindblad) the nonlinear theory only provides small corrections to the linear theory (this was done already by Vandervoort (1971)). But near the main resonances the nonlinear theory is not reduced to the linear theory in the lowest approximation. What happens is the same as the transition from the nonresonant form of the third integral to a resonant form.

There is an intimate relation between the density wave theory and the third integral and not only an analogy. In fact the density wave theory provides the distribution function f (which is the density in phase space) in the form of a series

$$f = f_0 + f_1 + f_2 + ... \qquad (15.1)$$

and the linear theory gives the form of f_1 if the axisymmetric term f_0 is known. The successive terms of this series become smaller as the order increases. The function f is an integral of motion, of the same form as the third integral. In particular, f_0 is a function of the unperturbed energy E_0 and angular momentum J_0: $f_0 = f_0(E_0, J_0)$.

But near every resonance the third integral is different. Near the main resonances even the first term f_0 is different. Thus, the form of the density distribution, near the Lindblad resonances and corotation, changes completely. E.g. the term f_1 is much larger than f_0 in this case. My first paper on this subject appeared in the Astrophysical Journal (Contopoulos 1970a). But later there was a series of systematic papers on the nonlinear density wave theory (Contopoulos 1975a; Contopoulos and Mertzanides 1977; Contopoulos 1979,1981a).

After that whenever C. C. Lin was speaking about the density wave theory and someone asked a question about nonlinear effects near resonances he would answer "Ask Contopoulos about that".

A problem of great interest at that time was why trailing waves are preferred, rather than leading waves. In fact a density wave theory can be applied to both leading and trailing waves and this was noticed already by B. Lindblad. C.C. Lin hoped that a higher order epicyclic theory might give preference to trailing waves, but this hope was not realized. In fact, only near the resonances (especially the inner Lindblad resonance) one can differentiate between trailing and leading waves. Then I worked out the response, near the inner Lindblad resonance, of a slightly growing wave. The wave was assumed growing, in order to avoid a singularity that appears at the inner Lindblad resonance in the case of a stationary wave. It was found that both in the case of a trailing and of a leading wave the response is trailing. This gives a clear preference of trailing waves.

This fact was already implied in the thesis of Agris Kalnajs, but in a few words only, that were not very clear. Agris Kalnajs explained to me what he meant. Then I presented the thesis to Alar Toomre and told him: "Kalnajs proves the preference of trailing waves in this particular page. Can you understand the proof?" Next day Toomre told me that he couldn't. Thus, I decided to write down my own results, that were published later (Contopoulos 1971a).

Around that time (1969) started a bitter controversy between Lin and Toomre, in which Shu and Kalnajs participated. Everyone was telling me that the others were wrong. Thus, I decided to organize a confrontation. The opportunity was given at the IAU Symposium No 38 in Basel during the summer of 1969 (Fig. 15). I asked all interested people (Lin, Shu, Toomre, Kalnajs, Lynden-Bell, Vandervoort) to attend an informal evening panel to express their opinions. There was not a convergence of the various views, but at least they were presented, side by side, in front of all interested people. So the differences of opinion were made clear. At the end of the meeting someone asked: "What was the purpose of this meeting?", and Toomre replied: "It was just to educate Contopoulos".

At another meeting (Cambridge 1986) I was the Chairman of a session, while Toomre was speaking. Toomre gave a nice talk about the regeneration of spiral density waves. But during his talk he made some jokes and insinuations about

Figure 15. At the Basel Meeting (1969). Sitting on the grass: G. Contopoulos, M. Hénon, A. Toomre and D. Lynden-Bell.

other people, which made the initiated participants laugh. At the end of his talk I thanked him and I added: "I suggest to Dr. Toomre to add a few footnotes when he publishes his paper, that should contain his jokes *with translation*". I am sorry that he did not follow my advice.

It seems that the inclusion of jokes in a talk is something rather common among American speakers. But such jokes are usually said in a slang language and in many cases they refer to situations that only a few "initiated" persons know. Non-Americans have great difficulty in understanding these jokes. I remember an international meeting, where an American was speaking in his usual joking manner. Then Dr. Heckmann (then president of the IAU) came to me and said angrily: "Tell your American friends to learn to speak English".

When we had to leave Boston (March 1969) I went to Chicago to find a house. Our house in Boston was rather small and, as we had two small children, we could almost never enjoy a peaceful sleep. In Chicago I went around with a friend looking for a larger house. We did find a big house, close to the University of Chicago. I called my wife and told her that we found a really big house, and she was happy. But then my friend wanted to make a joke. He added on the phone: "The only drawback is that there are black people everywhere around the house. And opposite us there is a famous black gang (not true)". My wife was terrified. She said that she would never come to Chicago under these conditions.

When I returned to Boston I explained to her that my friend was joking. But my wife had other reasons not to come to Chicago. She did not want to live permanently there, and she wanted to let the children grow up in our country. Thus, I decided to refuse the offer to take a permanent position in Chicago, which was very good from many points of view.

At any rate we arranged that my wife should return to Greece with the children, and I should go to Chicago alone.

My stay in Chicago was very profitable from a scientific point of view. Chandrasekhar had moved to Chicago, but I did not work with him on relativity theory, although we had close contacts and discussions on relativity and on many other subjects. I gave a course on dynamical astronomy, with emphasis on the density wave theory. I organized also a series of seminars with Professor Leo Kadanoff, on order and chaos in dynamical systems in general, and we both gave a series of lectures there.

Leaving Chicago I went to the IAU Symposium No 38 in Basel, Switzerland, on "The Spiral Structure of our Galaxy", which was held on the occasion of the 50th anniversary of the IAU. L. Woltjer was the president of the organizing committee.

The Symposium was very successful and covered both theory and observations. Bart Bok, in his closing lecture said: "We are fortunate indeed that the theorists attended the Symposium in force. The theoretical keynote address was brilliantly delivered by Contopoulos who gave us a full introduction to the gravitational approaches to the dynamics of spiral structure".

This Symposium was attended by 145 participants from 23 countries. Among the participants were W. Becker, A. Blaauw, B. Bok, G. de Vaucouleurs, J. Einasto, K. Freeman, O. Heckmann, C. Hunter, A. Kalnajs, F. Kerr, C. C. Lin, D. Lynden-Bell, S. McCuskey, R. Miller, W. Morgan, J. Oort, L. Perek, V. Rubin, F. Shu, E. Spiegel, A. Toomre, P. Vandervoort, L. Woltjer, etc.

Dr. W. Becker from Basel and I were the editors of the proceedings. We did a careful but strict work. We rejected one paper, that was irrelevant, and we changed considerably another paper by a Soviet astronomer that was full of attacks against his western colleagues. I wrote him that unless he accepted the changes that we had made his paper would be rejected. He replied that although he was protesting, he had to agree.

We had a problem with our Soviet colleagues. We had invited a number of astronomers related to the field. Instead, there came a group of people, most of which we had not invited. They brought me a theoretical paper by L. Marochnik, and asked me to present it. I told them: "Why are you not presenting it?" They replied that no one was a theoretician. I tried to read this paper, but its English was not clear at all, and I had difficulty in understanding its meaning. At any rate I presented a summary of it, and the proceedings contain only an abstract.

After Basel I was supposed to go to a meeting on "Periodic orbits, stability and resonances" in Sao Paolo, Brasil. They had sent me the ticket and they would cover all my expenses. But my wife told me, on the telephone, that she expected our third child any moment. Thus, I decided to cancel this trip. I gave the text to Michel Hénon who was kind enough to read it at the meeting. I cancelled also an excursion that I had planned, to Peru, to see the Matchu Pitchu. So I say, even now, to my daughter that I missed the Matchu Pitchu for her sake.

16. A VISIT TO ISRAEL

A friend of mine was Mike Lecar, whom I met at Yale in 1962, while I was a visiting professor there. Mike Lecar did some interesting work on a one-dimensional gravitating gas, and he presented it at the IAU Symposium No 25 on the theory of orbits, in Thessaloniki, Greece (1964). Later we collaborated in organizing together a number of meetings on the N-body problem, including the Besançon (France) Symposium on "New methods of stellar dynamics" (1966) and the IAU Colloquium No 10 in Cambridge, England on the "Gravitational N-body problem" (1970). At that time many people had started numerical calculations of the N-body problem. M. Lecar wanted to make a comparison of these calculations, done by different computers and different methods. Thus we wrote a letter to several people doing N-body calculations suggesting the same N=25-body system by giving the same initial conditions to all. The response was very satisfactory. Then Mike Lecar did a detailed comparison of the various calculations and he published his results in 1968 (Lecar 1968). The individual orbits of the stars deviated considerably from one calculation to the other, but several average quantities like the density function were rather similar. However it was rather unexpected that the number of escaping stars after a given period of time was not the same in different calculations but varied from 1 to 5. This result was not satisfactory, but Lecar was an optimist and he simply stated that the number of escapes was $N_e = 3 \pm 2$.

After these first attempts the number of separate N-body calculations increased considerably and to-day several calculations with N going up to many millions of stars have been performed.

Lecar was visiting Israel on a regular basis. In 1970 he arranged that I should receive an invitation to give a lecture on the third integral in Tel-Aviv. The invitation came from the head of the department, Dr. Yuval Neeman. Thus I met Neeman in Tel-Aviv for the first time. He was a quite remarkable person.

Neeman is well known for his early work on elementary particles. He developed a theory very similar to the theory of quarks of Gell-Mann, even before starting his graduate studies in England. But later he became deeply involved in politics. He was the leader of a very conservative Israeli party, which had some influence in Israel for some time.

When I met him I was impressed by his breadth of knowledge. He knew not only physics and several neighbouring branches of science, but also history and literature. He knew very well many ancient Greek authors.

But I was surprised by his views on religion. While he professed to be devoted to Israel, he was deeply anti-religious. I asked him: "How can you survive without your religion? Your religion is the link that binds Jews together, all over the world". "We are doing an experiment" he said. "I believe that we can base our tradition on our history and culture, without religion". I told him that this was a very doubtful experiment and it could fail.

While I was in Tel-Aviv I had lunch in the student cafeteria of the University. I met several graduate students that were half-soldiers. They served in the army part of every month and continued their studies the rest of the month. Some of them had just returned from a raid in Egypt, where they dismantled and brought back a russian-made radar. I asked their opinion about peace in the Middle East. They were very contemptuous of their adversaries. I told them that even if they succeeded in several wars, the Arabs nevertheless outnumbered them so much, that Israelis should be more moderate in the future. Their reply was really arrogant. "In the future we will have the atomic bomb" they said. Such an arrogance is very ominous for the future of peace.

17. FIRST IAU REGIONAL MEETING

My next visit to the U.S. was in 1971, as a visiting professor at the University of Maryland. There I gave a course on "the dynamics of spiral structure", that was published in a mimeographed form by the University of Maryland ("Maryland Notes", Contopoulos 1972). I had many useful contacts, especially with Frank Kerr and Alex Dragt.

Frank Kerr invited me several times to his house. On one occasion his wife (Maureen) at my insistence told me her story about the time she was a prisoner in a Japanese camp in Indonesia. It was an incredible horror. I admired Maureen very much for her courage, that enabled her to survive this death camp, and also her courage to remember such terrible things and relate them peacefully.

We had many discussions with Frank Kerr on various applications of the density wave theory to our Galaxy. I was preparing then a rather long paper on the corotation resonance in galaxies, which was published later in the Astrophysical Journal (Contopoulos 1973b).

At the same time I had many discussions with Alex Dragt, on integrals of motion, and their applications to the Störmer problem (motions of charged particles in the Earth's magnetic field) and to the dynamics of large accelerators. I was very much interested in these problems and, later, I published myself a paper on this subject, dealing, in particular, with resonances in the Störmer problem (Contopoulos and Vlahos 1975).

The "Maryland Notes" was the first exposition of the theory of spiral structure in book form. Although it was not a real book, nevertheless it attracted much interest and it was used by several people. E.g. Alar Toomre, who is not an easy critic, wrote me on 20-07-1972: "I have read them now from cover to yellow cover, and promptly write to applaud you for the general clarity and fairness of your exposition".

During 1972 we organized the first IAU Regional Meeting in Athens. The Scientific Organizing Committee, appointed by the IAU, consisted of A. Blaauw (chairman), A. A. Brück, G. Contopoulos, Ch. Fehrenbach, L. Gratton, P.O. Lindblad, E.R. Mustel, L. Perek and H. H. Voigt.

There were about 140 participants, mainly from Europe, from all branches of astronomy. The proceedings were published by Springer Verlag in 3 Volumes (a total of 810 pages). We had many important contributions, including some from leading astronomers of the world. Among them was S. Chandrasekhar, who emphasized that he did not go to any meetings normally, but he made this exception for the sake of his Greek friends.

But at the same time there was political opposition against our meeting, especially in France, because of the Junta. There was a slogan: "Do not go to Greece, while the Junta is in power". When I learned about that, I wrote a letter to J. C. Pecker, a good friend of mine, former General Secretary of the IAU and president of the French Astronomical Society at that time: "If you do not come to Greece the Junta will not suffer any harm. But you should come to support Contopoulos and the other Greek astronomers, who are under harassment from the Junta". Pecker was convinced. He not only came to our meeting, but he also encouraged others to come.

Then we had a serious problem with the minister of education. We had received financial support for our meeting from the Greek government. But when the minister saw the printed program of the meeting he was furious, because there was no mentioning of this support. He invited me to his office and told me that we should withdraw the programs and print new ones stating that the meeting was sponsored by the IAU and the Greek government. I replied that the IAU was at a different level and could not be a co-sponsor together with a government, or a local organization. "Then, he said, you have to cancel the meeting altogether". This discussion was taking place two days before the meeting started, and people were expected any moment. I replied that I could not take any initiative unless I consulted the president of the IAU. The minister insisted. Then I proposed a compromise. I said that we could write: "Organized by the IAU, with the financial support of the Greek ministry of education". This solution was reluctantly approved by the minister, and the meeting run smoothly after that.

We had invited to this meeting several Soviet representatives, promising to cover all their travel and living expenses, but only four Soviet astronomers were allowed to come. Nevertheless the fact that some Soviet people came, gave a good excuse for several Eastern European countries (Poland, Czechoslovakia, Romania and Bulgaria) to send many of their astronomers to our meeting. In particular there were many people from Romania. We were providing free lodging for all of them, and we expected them to provide their own meals. But they were given only $1 per day from their government and it was impossible to survive in this way. Thus, we found other ways to help them. Furthermore we organized some receptions for all participants (Fig. 16).

Figure 16. At a reception during the IAU regional meeting in Athens (1972) (Mrs. Contopoulos, Mrs. Chandrasekhar, Dr. Oort, G. Contopoulos and Dr. Chandrasekhar).

At any rate the meeting was quite successful and it marked the beginning of similar regional meetings in Europe, Asia, Latin America, and even in Australia (in North America they had already the meetings of the American Astronomical Society).

After the Regional Meeting I took S. Chandrasekhar and his wife and drove to Thessaloniki where they spent a few days. We went first to Delphi and then we continued driving past Mt. Olympus, during the evening.

We had in the car my small daughter, who could not sleep and wanted to hear a fairy tale. Chandra volunteered to tell her a story. He spoke about the gods of Himalayas, and I was translating. At the end Chandra asked my daughter if she liked the story and she replied: "I did not understand a word of what you said. But what my daddy said was very nice".

The day of Chandra's departure there was a strong wind and all the planes to Athens were delayed. Chandra was unhappy because he had to catch a plane for Vienna. When finally the planes started to fly I asked the authorities to put Chandra and his wife in the first plane. But after they were on board I realized that they had not transferred their luggage also. So I rushed to transfer their luggage the very last moment.

Chandra related his part of the story as follows (letter of 2-10-72): "Lalitha and I simply could not believe our eyes when we saw you from the plane, a few minutes before it took off, standing outside directing the porters to put our baggage in the plane in which we were and not in the other one. Before we got on the plane we knew that our baggage was on one of the carts in one of the lines. And Lalitha and I were anxiously watching whether the baggage on that cart would be transferred to our plane or to the one standing next to ours. And when we noticed that it was not going to happen, our hearts sank. And suddenly, lo and behold you were there directing our baggage to our plane. We have never witnessed a miracle like that before! And you can imagine how grateful we were. The four days we had with you and Vaya are among the four most beautiful days we can recall anywhere. The unexpected strong winds from Mt. Olympus, I am afraid, must be traced to the wrath I must have produced in the Gods on Olympus by extolling in their presence the virtues of the Gods of the Himalayas!".

18. CONTACTS IN ITALY

After the Athens Regional Meeting I was invited by the Italian Physical Society to give a lecture at the Cagliari Meeting of the Society. Two professors, Galgani and Scotti, wanted to hear details about the third integral.

I flew to Rome, and then to Cagliari, in Sardenia. Galgani and Scotti were waiting for me at the airport of Cagliari. We went to the hotel and started our discussions. We finished late at night, and we continued next morning. Then we had a quick lunch and after that I had my lecture and further discussions, until it was time to leave for the airport. I had no time to see Cagliari, or Sardenia, which seems to be a very nice island, except during our driving to the airport.

But our discussions were very fruitful. Scotti and Galgani had worked on the Fermi-Pasta-Ulam problem (a system of many particles along a line). This is called also the "Fermi-Pasta-Ulam paradox" (section 12.3). The reason is that Fermi et al. (1955) expected such a system to be chaotic, because of the non-linear interaction between the particles; in particular they expected an equipartition of the energy among the various modes of oscillation. In order to check that they performed one of the very first computer calculations (1955). To their surprise there was no equipartition at all, but the energy was shared between only a few modes, in a periodic way.

The reason of this non-equipartition is the existence of integrals of motion, like the third integral. But these integrals are only formal and if the energy is sufficiently large, there is a complete exchange of energy, leading to equipartition.

As in other problems of the third integral type, the transition is rather abrupt. If the energy is below a critical value the sharing of energies is very small, while if it is above this value the sharing is complete, and equipartition is established.

The main question of Scotti and Galgani was what is the critical energy, and how it is related to the number of particles (or degrees of freedom). Some numerical calculations of Scotti and Galgani indicate that the critical energy per particle E_c/N is constant for large N. On the other hand, other people (e.g. Chirikov, and Froeschlé) hold the view that the critical energy decreases to zero if N is large, e.g. larger than 4.

An argument in favour of the last view is the fact that the various parts of the phase space can communicate if the number of degrees of freedom N is equal or larger than 3. Then we have a diffusion , that is called "Arnold diffusion", throughout the whole phase space. But Arnold diffusion is extremely slow, as it was found by numerical experiments, and also by theoretical considerations (Contopoulos, Galgani and Giorgilli 1978). It requires exponentially long times, or even superexponentially long times (Morbidelli and Giorgilli 1995) to cover the whole phase space. Therefore, equipartition may occur after such long time intervals, that it is irrelevant for most problems of physics and astronomy.

After my first contacts with the Italian group, a student of Galgani (A. Giorgilli, later professor at the University of Milan) did his thesis on integrals of motion, and became later a leading person in this field. We had many contacts with Giorgilli and Galgani over the years and we continue to publish some papers together.

Galgani extended his research in a different direction also, namely a classical alternative to quantum mechanics. If there is no equipartition of energy among coupled oscillators, one may derive the Planck black-body spectrum by classical considerations. In fact a classical derivation of Planck's law was already given by Einstein and Stern (1913). Further results of quantum theory, like the zero point energy and the specific heat of a gas, were derived classically by Galgani and his associates. Galgani introduces a classical constant of action, h_c, corresponding to the Planck constant h. The value of the classical constant is of the same order of magnitude as the Planck constant h, but we have no proof that it is exactly constant. Thus, the whole classical theory of quantum phenomena is still ambiguous.

More recently Galgani and his associates attacked the problem of non-locality of quantum mechanics, by using limiting forms of the classical electromagnetic theory, including the Abraham-Lorentz-Dirac equation of motion. This equation uses not only second derivatives with respect to the time but also third derivatives. Because of that its solutions require some nonlocal restrictions. Thus, it may be possible to bypass the "Bell inequalities" that eliminate all local theories of hidden variables. Whether nonlocal theories, like the one based on the Abraham-Lorentz-Dirac equation, are correct, only the future will show.

19. SERVICE IN THE IAU

At the Brighton General Assembly of the IAU in 1970 I was elected Assistant General Secretary of the IAU. My main duty was to supervise the various Symposia and Colloquia of the IAU (about 50 of them). In several cases I had to attend various meetings myself (Fig. 17). But what was most important was the establishment of the Regional Meetings of the IAU, starting with the 1st European Meeting in Athens, during 1972. This effort was really successful. Up to now (2004) there have been 30 Regional Meetings in Europe, Asia-Pacific and Latin America[6].

I had to attend also the meetings of the Executive Committee (one per year) and the meetings of the officers, namely of the President (B. Strömgren), the General Secretary (K. de Jager) and the Assistant General Secretary. We had three meetings every year, one in Copenhagen, one in Utrecht, and one in Thessaloniki.

In Copenhagen we enjoyed the hospitality of B. Strömgren, who was the "first citizen of Denmark", at his sumptuous mansion. This house was a donation to the Government by the Carlsberg breweries. It was vast, with many paintings and sculptures everywhere, and a magnificent flower garden, attended by four gardeners. The Carlsberg foundation provided also free beer. Besides an ample collection of bottles they had a tap, that would provide beer instead of water.

The first time that we went there we had some difficulty in finding the mansion. As no one spoke Danish we asked some people in English, French and German, with no success. Then de Jager said: "I will try Dutch". That was successful. It seems that Dutch and Danish are rather close together.

In Utrecht we had the secretariat of the IAU, and we could take care of many business matters rather fast.

[6]The European Meetings after 1990 continued as Meetings of the European Astronomical Society.

Figure 17. At a meeting in Oporto, Portugal (1970), with R.G. Giovanelli from Sydney.

In Thessaloniki I was the host. But I tried also to get some profit from the visits of such distinguished guests. On one occasion I asked Dr. Strömgren to give a lecture to my students. As most students did not speak English, I was translating for them. (A translation is not necessary in recent years; on the contrary it is very common that the speaker changes his language from Greek to English if he realizes that a foreigner is present in the audience).

Strömgren gave a nice lecture, but sometimes difficult for the students. Then I had not only to translate, but also to add some more explanations. At the end I thanked Dr. Strömgren, and he said, smiling: "I noticed that you improved considerably my talk".

After three years as Assistant General Secretary, I was elected General Secretary at the IAU General Assembly of Sydney, Australia (1973)(Fig. 18). At the same meeting Leo Goldberg (Kitt Peak) was elected President and Edith Müller (Geneva) Assistant General Secretary. We continued our meetings of the Officers and of the Executive Committee of the IAU, both in Europe and in the U.S.A. Everyone was happy when we had meetings in Greece.

On one occasion I had organized an excursion to our new Observatory in Kryonerion, near Korinth. I had my youngest daughter with me, who was then learning French. I told her that Mrs Edith Müller was French (Swiss) and I suggested: "Go to speak to her". My daughter tried to speak to Edith, but she came back disappointed. "She doesn't speak French" she said. "I asked her how we say a house in French and she did not know that it is maison". What

Figure 18. With the new President of the IAU, Leo Goldberg in Sydney (1973).

happened is that my daughter asked Edith in Greek and wanted her to reply in French!

The job of the General Secretary was very different from that of the Assistant General Secretary. I had to face many problems of political, rather than scientific, nature.

One of our first efforts was to bring China back to the IAU. China had withdrawn from the IAU when Taiwan was accepted as a member country. We sent many letters to Beijing, without apparent success. But, somehow, the relations were improving. At first the Chinese wanted the IAU to expel Taiwan from the Union. But later an unexpected compromise was found. China asked us to state simply that there is only one China, but there are two Chinese delegations in the IAU, one from the mainland, and the other from Taiwan. When this was accepted and the delegation from Beijing came to the IAU, the head of the delegation thanked me and the other IAU officers for our efforts to bring China again into the IAU.

Another serious problem, that we faced already in Brighton, was the place of the 1973 General Assembly. There were two invitations, one from Australia and the other from Poland. The Polish invitation was based on the fact that the year 1973 was the 500th anniversary of Copernicus.

The Executive Committee had already accepted the Australian invitation, but the Soviet Union backed the Polish application and threatened to boycott the General Assembly in Australia. Thus, it was decided to have an Extraordinary General Assembly in Warsaw.

The Extraordinary General Assembly was the responsibility of the new Presidium. I had been already in Warsaw to supervise the preparations for this General Assembly during 1972.

I remember that I had to fly to Frankfurt and then take a plane from Frankfurt to Warsaw. But my plane was late in arriving to Frankfurt. The departure gate for Warsaw was already closed, but they told me that the plane had not yet left. Thus, I made a desperate, but foolish, move. I went to the runway with my luggage and I tried to reach the plane. But I was soon stopped, and forced to go back to the building.

Then I tried to find another solution. The only possibility was to fly to Budapest and take a connection to Warsaw. But they could not give me an o.k. for the flight Budapest-Warsaw. Nevertheless I decided to go.

When I arrived in Budapest I realized that there was a mess (after that time the Budapest airport has improved dramatically). First of all it seemed that no one was speaking English. My German was bad, but nevertheless I could speak with someone on the phone, who said that I was on the waiting list. Then I had to wait. From time to time they were telling me that the plane was full. I had not enough money to go to a hotel (because all my expenses were paid beforehand) and next morning I had some important meetings in Warsaw. I was very unhappy. I only remembered that my small daughter was praying that I should not be lost in a foreign airport.

It seems that the prayers of my daughter were successful, because the last moment they told me that I would be admitted in the plane to Warsaw. I was greatly relieved. When I entered the plane I noticed that many seats were empty. I asked why and they told me that the seats were reserved for some high ranking authorities, that cancelled their trip the last moment.

I arrived very late in Warsaw, but that did not matter. They only told me that I missed a ballet theater performance, for which they had bought a ticket for me. Then they took me to my hotel. When I entered my hotel room, I was amazed. That was not really a hotel room, but a large apartment. It had two bedrooms, a working office, a patio with large trees and decorations, televisions, etc. They explained to me that this was normally reserved only for heads of state.

We came back to Warsaw, after the Sydney meeting (1973). We had a very long and tiresome flight. But our Polish colleagues made every effort to satisfy us. The night of our arrival they gave us some gifts, among them a facsimile of the original manuscript of Copernicus. There one can notice a page where Copernicus refers to Aristarchos of Samos and his heliocentric theory. But this page is crossed out in the manuscript and it did not appear in the published version of the "De Revolutionibus".

The day after our arrival we had the official opening of the Extraordinary General Assembly. It was very solemn. I noticed that the minister addressed his audience as "dear comrades", but the organizers addressed the minister as "citizen minister" and not as "comrade minister". I thought that this was a mark of defiance. But they explained to me that they did not belong to the party, and thus they were not allowed to use the "prestigious" expression "comrade".

The speakers emphasized, again and again, the heliocentric theory developed by Copernicus. I was sitting next to the President of the IAU, L. Goldberg, and I showed him the page of the manuscript of Copernicus, where he referred to Aristarchos. Goldberg made a note of that. When he got up to speak he started by addressing the authorities, and the organizers, and then he added: "dear fellow-travellers from Australia", a jocular expression that was completely out of tune from the style of the local people. Then he mentioned Copernicus, and made the following remark: "Copernicus was very careful in making the proper acknowledgements. In the manuscript of his book he refers to Aristarchos of Samos, who was the first to propose the heliocentric theory".

The Warsaw General Assembly attracted many people from Europe, that could not go to Australia. It was a very good meeting, and it was followed by a number of Symposia in Warsaw, Cracow and Torun. I went to Cracow to attend the IAU Symposium No 63 on "Confrontation of cosmological theories with observational data" by bus. I noticed a person in our bus that was on a wheelchair. I was surprised that a person like that could travel so far away. Then they told me that this person was Dr. Hawking from Cambridge. We talked a little, but only through an interpreter, who could understand his words (later on his voice failed completely, and he was obliged to use an artificial voice). I always wonder how he could work with so complicated mathematical formulae without being able to use his hands.

In Cracow I met also for the first time Roger Penrose. We had some interesting discussions on cosmological problems. Penrose remarked that the entropy of the Universe increased to its present value (10^9 per particle) at the very early Universe. But if the Universe will recollapse, its entropy will reach a value 10^{40} per particle. Therefore the final state of a collapsing Universe is very different from its initial state.

The opinions of Penrose differ from those of Hawking. A very interesting confrontation between the views of Hawking and Penrose was published in the form of a book "The Nature of Space and Time" (1996). I must say that I agree with Penrose in most (but not all) cases.

Among other things, Penrose proposed that the ultimate Theory of Everything, that would unify gravity with the three other forces of nature, should be based not on a quantization of gravity, but on a nonlinear quantum theory, in which the nonlinear effects would be of gravitational nature.

We discussed with Penrose the possibility to formulate such a theory on another occasion. But it is too early to say whether such a theory would be successful.

In Cracow we had programmed an invited lecture by Dr. Ambartsumian, former president of the IAU. But when Ambartsumian arrived, he had his lecture only in Russian. He asked for an interpreter. But there was no one who could translate from Russian to English. Our Polish colleagues then found two interpreters. One from Russian to Polish and another one from Polish to English! But a double translation would take such a long time that it was impossible. Then Ambartsumian said that he would translate his lecture himself, and he would speak in English. He spent the morning for that purpose, but his translation was not good and his pronunciation worse. Most of the audience were literally sleeping during his talk. Thus, I decided to do something daring. After his talk I got to the podium and said "I will explain to you in a few words what were the main points of the lecture of Dr. Ambartsumian". I spoke for less than 5 minutes, but the atmosphere changed. At the end people clapped hands enthusiastically.

I had in general good relations with Ambartsumian. When I first met him, in Brighton (1970), he was very much in favour of forming a European Astronomical Society (but his authorities did not give him the permission he wanted). We met one more time after the Polish incident. Late in 1973 there was a meeting of ICSU (International Council of Scientific Unions) in Leningrad, and Ambartsumian was the organizer. I was representing the IAU at that meeting. There was a lot of discussion about international collaboration in science, and free circulation of scientists. But Ambartsumian gave a chilling talk. He said that international collaboration should be scheduled between countries (internations) and not between individuals. Therefore, he objected completely to our practice of inviting individual Soviet scientists to our meetings. He said that our invitations should be sent to the Soviet National Committee, and they should decide who was to come to any particular meeting. After that no one wanted to speak. But I raised my hand. I said that Greek scientists suffered a similar isolation by our regime (it was then the Junta period), and the results were very bad. And I added: "No country can afford to isolate its scientists from the rest of the world, no matter how big this country is". I am sure that Ambartsumian was forced to say his comments and he did not believe them. In fact, despite my open disagreement with him, our personal relations remained always friendly.

I take this opportunity to relate my trip home after the Leningrad meeting. I flew to Moscow alone, without any escort, and I had instructions to go to the Aeroflot hotel. There was a bus going there from the airport. That was fine. But then I had to return to the airport very early next morning. At the reception they gave me a note for a lady that was taking care of the third floor, who would give me the key of my room. When I got the key I tried to explain to her that she should order a taxi for me at 5 o'clock in the morning, and she should wake me up at 4:30 a.m. She said that she did not understand English, French, German, or any other language except Russian. So I went down to the reception, and the only thing they told me was: "But it is her duty to help you". I protested: "How can I tell her what is her duty if she does not understand me?" and the answer was: "If you give her two roubles she will understand". (At that time one dollar cost officially 0.60 roubles!). I gave her two roubles, and she started speaking very fast in Russian. I hoped that she understood, but I was not sure. So I did not sleep at all, afraid that I might miss my plane (I had no alarm clock with me). I got up early and I was at the entrance at 5 o'clock. But no taxi came. I tried to speak to the person at the reception but he spoke only Russian. Finally at 5:30 a group of Japanese tourists boarded a bus for the airport. I entered the bus with them. At that moment came the lady of the third floor shouting that I should go out because the taxi was coming. I refused to go out. Finally, the Japanese tourists protested and the bus left. I just made it in time for my plane.

20. IAU SECRETARIAT

After the Polish General Assembly (1973), we established the IAU Secretariat in Thessaloniki. Two devoted secretaries, A. Jappel and J. Dankova, both from Czechoslovakia, moved to Thessaloniki and helped me considerably in organizing the office of the IAU. The work was heavy, and our correspondence reached some thousands of letters every year. But things run smoothly, despite some difficulties during the Junta period.

One unexpected difficulty was related to my contacts with the President of the IAU, Dr. Leo Goldberg. I was used to the previous regime, in which the President, Dr. B. Strömgren, left everything in the hands of the General Secretary, Dr. K. de Jager. Even the letters of the president were prepared by de Jager, and sent to be signed by the president. In fact according to the statutes of the IAU the General Secretary is the only legal representative of the IAU.

Now the situation was different. Dr. Goldberg wanted to participate in all the activities of the IAU. He had strong opinions about many things and we disagreed on a number of points. As a third person said: "The Strömgren era was like a European democracy, in which the president is a figurehead and the prime minister has all the power. But Goldberg wanted to be an American-style president, who has all the power in his hands".

I did not want to solve our differences by legalistic discussions, and I tried to argue politely with the President. But in a few cases I had to insist. Finally a modus vivendi was established that reduced the friction considerably. I am happy that L. Goldberg, in his closing speech as President of the IAU in 1976 acknowledged my efforts to work in harmony with him (see section 23).

After that we became close friends, and our relations were really relaxed. Leo Goldberg used to call me around noon from his office in Tucson, Arizona. He would start with a friendly "Good morning George", but because of the 9 hours difference between Tucson and Greece it was night in Thessaloniki and

I was just going to bed at that time. Leo Goldberg supported me strongly in some difficult situations, one of which I will describe now.

During 1974 there were some serious developments affecting Greece. First of all there was a second dictatorship in our country. A particular Saturday night some lower ranking officers arrested the former dictator Papadopoulos, and formed a new Junta.

The day after the coup d'état was a Sunday. We did not know anything, and I was driving my family to the church, when we were stopped by the police and turned back. The same day Jappel and Dankova had decided to go on an excursion. They started with their car, but they noticed several tanks in the streets. They thought that there were some military manoeuvres taking place. Thus, they went quite a distance, until they were stopped by an officer with a group of soldiers. Jappel tried to explain that they were going on an excursion. Then the officer said in German: "Verboten". "Why verboten"? asked Jappel, and the officer explained in bad German: "Wir haben Revolution. Papadopoulos kaput. Andere Papadopoulos, andere Demokratie" (We have a revolution. Papadopoulos is down. There is a new Papadopoulos, a new democracy!).

But the "new democracy" (i.e. the new Junta) failed completely. After a few months, in order to avoid the mounting opposition of the Greek people, they tried to rally the Greeks by staging a coup in Cyprus that overthrew the President of Cyprus, archbishop Makarios. But then Turkey intervened (1974). The coup was not directed against the Turks, but nevertheless Turkey found an excuse to invade Cyprus, claiming that they wanted to re-establish the status quo. But in fact the Turkish occupation of the northern part of Cyprus continues until today.

The Turkish invasion brought the collapse of the Athens Junta (1974). Many military realized that Greece was going to collapse and they decided to bring back the politicians to govern the country. Thus, they invited Caramanlis, a former prime minister, who was in exile in Paris, to take over. Caramanlis was welcomed by millions of Greeks, when he came back, but had to face extremely hard problems. He asked the help of NATO. But NATO tried to avoid any involvement in the Cyprus crisis and Caramanlis withdrew from the military branch of NATO. Finally an agreement to a cease-fire was reached with the Turks, but 38% of the island are still under Turkish occupation, despite the decisions of the United Nations.

During the Turkish invasion of Cyprus all flights to Greece were cancelled and all telephone links were interrupted. I was trying desperately to contact my family or to find a flight to Athens, without success.

Only later I learned what happened with my family during these days. My wife had scheduled a transfer to a new house the very day of the invasion, and some of my assistants had volunteered to help her. But when all the furniture was loaded on a big truck and the truck moved to the new house, they heard on the radio that there was a mobilization of the Greek army and all reservists had to present themselves immediately at the mobilization centers to be moved to the front. My assistants had left already. Even the driver of the truck had just time to unload the furniture in the street and rushed away. Thus, my wife was left alone, with four small children, to move everything to the 9th floor of an apartment house. I still wonder how she managed to do that.

A few months later there should be the General Assembly of ICSU (International Council of Scientific Unions) in Turkey. I was supposed to represent the IAU at this meeting. But the Turkish authorities did not want to give me a visa. Then Goldberg protested to ICSU and to some of the leading Academies of the world. One of the main principles of the ICSU is the free circulation of scientists. It was clearly stated in its byelaws that no meeting should take place in a country that does not allow all foreign participants to attend such meetings. But the ICSU president did not want to take such a drastic decision and cancel the ICSU meeting in Turkey, for which so many preparations had been made already. He suggested (informally of course, but I heard it myself) that the IAU should send another representative. But Goldberg was firm. "Only Contopoulos will represent the IAU" he said. At the same time the National Academy of the USA, the Royal Society of England, and the Swedish Academy of Sciences protested strongly to the Turkish authorities. I learned later that some Academies had instructed their representatives to walk out in protest if Contopoulos would not come to Turkey.

Finally the Turkish authorities gave in. The very last moment they informed me that I should go to the Turkish consulate in Thessaloniki, to get my visa. It was a Saturday and the consulate was closed. Nevertheless, I got my visa and arranged immediately my travel.

The meetings took place in Istanbul and Ankara. We were invited to some receptions in some old palaces. They were impressively decorated for the occasion with many nice sculptures made on ice. But they did not serve any alcohol, and some people were angry.

In Istanbul I met an interesting person. That was the director of the Istanbul observatory, Mrs N. Gökdogan. Mrs Gökdogan was a Greek married to a Turk. She told me proudly: "I am a "romia" from Polis". "Romios" fem. "romia" means a Greek (derived from the Eastern Roman Empire), while Polis means "Constaninopolis", the city of Constantine the Great. In fact even the Turkish name of Constantinopolis, Istanbul, is a Greek expression meaning "to the city" (is tin polin).

21. VISITS TO THE USSR

After my first visit to Moscow in 1958, I visited the USSR again in 1973 to prepare the 1975 IAU Regional Meeting in Tbilissi, Georgia.

My Soviet colleagues were waiting for me at the airport. At the customs they were checking carefully the luggage of all passengers. But when I said "Akademia Nauk" they let me pass without opening my luggage at all. I used again these magic words when I visited St. Basil's church in the Red Square. The church was a museum, and when I arrived there it was closed. But when I said "Akademia Nauk" they opened it just for me.

In 1973 my colleagues at the USSR Academy had arranged for me a trip to the site of the 5m telescope in Caucasus (Fig. 19) on my way to Tbilissi. A young astronomer escorted me all the way. That was absolutely necessary, because I was not speaking Russian, and our trip was really complicated.

Figure 19. At the 6m Telescope in Caucasus (1973).

We were supposed to land at Minerali Vodi, north of Caucasus, where a car from the Observatory was waiting for us. But a fog at the airport did not allow us to land there. The plane crossed the Caucasus mountain and landed in Sukhumi on the Black Sea, until the weather cleared in Minerali Vodi. We arrived there after many hours, and the car was waiting for us. We travelled for about four hours to reach the Special Astrophysical Observatory, near Zelenchuskaya. The landscape was mostly barren with only a few cultivated areas near some villages. The villages were not like any villages in other countries. They consisted of

3-4 high rises in the middle of nowhere, without any small houses nearby. Only the larger cities consisted of large numbers of small houses.

When we reached the site of the 5m telescope we found a nice small hotel and a western-style cafeteria. The headquarters are in a nearby valley covered by a thick forest. In the neighbourhood there are the remains of some very old churches built in the early Christian centuries, that were restored only recently as museums (Fig. 20).

Figure 20. With Y.Y. Balega, director of the Special Astrophysical Observatory in Caucasus, in front of an old church near the Observatory.

The scientific environment in the Observatory was very good. I gave a seminar in English and a colleague was translating into Russian. When I thanked him I used the expression "you Russians" and he objected. "I am not a Russian, he said, I am Ukrainian". I had not realized until that time the deep diversification of the various nationalities in the USSR.

During my visit I had a long discussion with Parijskij who was in charge of the RATAN-600 radio-telescope nearby. Among other things he described his work on the 3°K "relic radiation". At that time it was not quite certain that this radiation was primordial, and some people believed that it was due to galaxies, or clusters of galaxies. I asked him: "The word "relic" that you use seems to imply that this radiation originated in the big bang. Do you believe in the big bang"? "Yes" he said. And then I tried to tease him. "How come, that you, who are coming from the country of dialectical materialism, accept the big bang, while others, in the west, make every effort to avoid any notion of big bang, as idealistic and religious"? He smiled and he said: "I also tried to ascribe the microwave radiation to galaxies and clusters. But when the accuracy of my observations reached a level ten times lower than the expected irregularities due to galaxies, and no such irregularities were found, I had to accept that this radiation is primordial. After all science is not what we wish, but what we observe". Very well put, indeed.

At the Observatory they told me that there was a Greek village nearby. I was surprised. I knew that there were many Greek villages south of Caucasus, in Georgia, especially near the Black Sea. But I had never heard of Greek villages north of Caucasus.

As we were leaving the Observatory we visited the site of the RATAN-600 radio-telescope, that was then under construction. My interpreter asked if there was any Greek among the workers. Someone came to us and spoke in Russian. I spoke to him in Greek and he was astonished. "Where are you coming from"? he asked. I told him "From Athens". He was so surprised as if I had come from Mars. They had never seen a person from Greece before. Then I asked him: "Are you married?" in modern Greek, and he did not understand. I repeated the question in ancient Greek and he immediately responded. In fact these people were using a dialect that was closer to the ancient Greek, than the Greek used in present-day Greece.

As I learned later these people had came to this area from Trapezous (Trabzon) in Asia Minor, escaping the Turkish genocide during the first World War (that resulted in more than 1.500.000 victims, 1.000.000 Armenians and 500.000 Greeks). They formed a village called "Hasaout Grecheskoe", i.e. "Greek Village on the river Hasaout", and they established a church and a school there. The church was demolished later by the communist regime and the school was closed, but the children learned Greek from their parents.

When we came back to Minerali Vodi we had to take an evening plane to Tbilissi. They allowed me and my escort to enter the plane first. Curiously enough the plane was completely dark. Nevertheless we found our places and then the local passengers rushed in. They came with a lot of hand baggages, including their poultry and other small animals. They were shouting in the dark, and their voices were mixed with the chuckling of the chicken. Finally the pilot came in. He put on the lights and we left.

In Tbilissi they took good care of me. I gave a talk to the staff of the Abastumani Observatory. But I realized that their knowledge of theory was very far behind. What was even more strange for me was their lack of contacts not only with the west, but also with the modern Russian institutes and also the Armenian institutes further to the south. I had the feeling that those people considered Moscow equally far away as New York. I realized that there was a profound division between the various nations of the Soviet system. This division led later to a complete separation of the constituents of the Soviet Union.

I came back to the Soviet Union in late 1973 to attend a meeting in Leningrad. On this occasion I visited the Institute for Theoretical Astronomy. I knew the vice-director, Dr. Victor Abalakin, who was later director of the Pulkovo Observatory. I met there a number of astronomers, among them a lady, Mrs. E. Kazimirtchak-Polonskaya. Dr. Abalakin was acting as interpreter. I felt somewhat embarrassed when Mrs. Polonskaya thanked me for sending her some Russian books on philosophy and religion from Paris. In fact she had asked me on a previous occasion to send her some books of Berdyayev, the well known philosopher, and Bulgakov, a famous Russian orthodox theologian and priest, émigré in Paris. At that time the circulation of such books in Soviet Union was absolutely prohibited. Nevertheless I had sent her these books through a third person, that was travelling to the Soviet Union. So, I felt a little uneasy to speak about these books. I remembered that at the Leningrad airport they had checked carefully all the books and journals that I was carrying.

After Mrs. Polonskaya left the office, Dr. Abalakin suggested that we should go out for lunch. When we came out in the street, Victor Abalakin told me: "I could not speak freely in the office, because there may be hidden bugs. But I wanted to tell you that Mrs.Polonskaya is a very brave woman. She suffered a lot because she professed openly her Christian beliefs. But they could not fire her because of her international reputation".

Several years later, when I visited Dr. Abalakin at the Pulkovo Observatory, where he was then director, he showed me a room dedicated to Mrs.Polonskaya, who had meanwhile passed away. Before her death she had become an orthodox nun.

In 1975 we had the 3rd European Regional Meeting of the IAU in Tbilissi, Georgia (USSR). We were invited by Prof. Kharadze, Vice-President of the IAU, and president of the Georgian Academy of Sciences. There were many people attending, especially from the USSR. I went there with my wife, and we had a very good time.

The people of Tbilissi were very hospitable. The taxi drivers, when they realized that we were coming from Greece, were very friendly and they practically never accepted to be paid! One driver explained to us why. He said "Greek = orthodox" and he made the sign of the cross in the orthodox way.

We were staying in the best hotel of the city (hotel Iveria) and several touristic guides were at our disposal. The main touristic attraction of Tbilissi were its churches. Old orthodox churches with a peculiar architecture, typical of Georgia. The oldest church dated from the 3rd century A. D. My wife became good friend with a young lady guide. We invited her to our hotel to tea, or dinner, but guides were not permitted to enter the hotel. My wife asked her if she planned to go abroad to continue her studies. At first she claimed that she only had to choose between England and USA. But the last day, as we were separating, she burst into tears and told us that she would never be allowed to go abroad, and she would spend her whole life in misery.

There is a funny sequence to that story. Next year we were at the Grenoble General Assembly and my wife wanted to send a gift to this lady guide in Tbilissi. She asked some young astronomers from Tbilissi, but no one dared to take it, because they were very afraid. Finally a lady astronomer said: "O.k. I will do that for you. But I do not want to be seen taking anything from you. So you will leave it in your pigeonhole and I will take it from there". We did not want to put into trouble this brave person, so we asked the head of the Soviet delegation himself, Dr. Kharadze, to bring our gift to our friend in Tbilissi. He was kind enough to do us this favour.

During the Tbilissi meeting I spoke about the places of formation of stars in the solar neighbourhood, describing mainly the work of Preben Grosbøl, who was doing his thesis with me at that time.

Among the participants was Per Olof Lindblad, and I was surprised when he told me that he would speak about the work of Grosbøl. In fact he had met Grosbøl in Copenhagen and he knew his work. But Lindblad's interpretation of this work was very different from mine. He assumed that star formation started at the shock formed by the gas when it entered a spiral arm (the maximum density of the density wave) and continued all the way to the next shock at the next spiral arm. On the contrary I assumed that star formation was taking place only close to the spiral arms.

This discrepancy led me, after my return to Greece, to ask Grosbøl to make some extra calculations to decide what view was correct. Grosbøl assumed various values for the rate of star formation away from the spiral arms and compared the theoretical results with the observed data. His conclusion was that star formation takes place mainly close to the spiral arms although not exclusively inside the spiral arms.

During this meeting we heard a very nice lecture by Shklovsky. He was an excellent speaker. He was also a firm opponent to the Soviet regime. For this reason he was constantly harassed. We invited him as an "invited speaker" at the IAU General Assembly in Grenoble (the highest honour of the IAU) but he was not allowed to come. At the General Assembly the Soviet delegates told us that Shklovsky could not come for personal reasons, and added that it was bad of him that he did not inform us. But Shklovsky had informed us in a different way. He sent a postcard with the remark that he was now working on black holes. And he had made a sketch of a black hole, with Shklovsky himself inside it! The meaning was obvious.

In Tbilissi I had a long discussion with Shklovsky as we were sitting outside the meeting room. He spoke in such an outspoken way about the regime, that I felt embarassed. People around us were taking photographs continuously. But he did not care.

After Tbilissi we travelled to Moscow and then to Athens. When we entered the plane to Athens we were surprised hearing several people speaking Greek. They explained to us that they were among the children that had been taken out of Greece during the civil war and they lived for several years in Tashkent, Uzbekistan. They had grown up, were married, and now they were allowed to return to Greece. I asked them many questions about their life in Tashkent but they were very reserved. After a few hours of flight we entered the Greek air space, and passed over a big city. I explained to them that this was Thessaloniki, and only then they started speaking openly. They told us what they suffered in the Soviet Union and how happy they were to return to Greece.

Next to us was a family with two small children. They had to travel to northern Greece by train the same evening. But they had very little money. So we suggested that they should come with us. We went to our house for a few hours of rest and then we drove them to the train station and we provided some money for their train tickets. They were so grateful that they wanted to give us a gift from their extremely meager baggages (they had to leave most of their belongings when they left the Soviet Union). We were quite moved by their gesture but we could not accept any gift.

Our relations with the Soviets were sometimes difficult. Goldberg and I had sent a letter of protest when a Soviet astronomer was arrested and exiled to Siberia. We had also expressed our support for Zakharov, who was restricted in Gorki. After that I had a visit from Mrs. Massevich, a high ranking official among Soviet astronomers. She came to Thessaloniki and gave a lecture. Then I took her out to dinner. During the dinner, without any obvious reason, she started attacking Zakharov and his followers, who were not "patriotic" and "circulated slander about the Soviet Union". I asked: "Why do you tell me all that"? She replied: "It is unbelievable that the President and the General Secretary of the IAU can take such a polemic position against our country. What do you want? To cut all relations with the Soviet astronomers"? I let her speak for a long time. Then I said: "Now I understand what was the purpose of your visit". She protested weakly. I told her: "You were asked to tell me your official point of view. You told me. Now tell "them" the following: As General Secretary of the IAU I feel an obligation to all astronomers of the world, including all Soviet astronomers, both those loyal to your regime and the dissidents". I mentioned our efforts to give financial support to Soviet astronomers to attend meetings etc, and I finished: "I act as a fellow astronomer, not as a politician. But if you put politics above science, you do harm to yourselves".

On another occasion I had a long discussion with Khromov, another leading figure in the Soviet astronomical bureaucracy. We had a long walk in the streets of Moscow, lasting well beyond midnight. He was thoroughly candid. He explained to me the basic principles of the Soviet system of marxism-leninism. It was the best crash course I heard on this topic. And then he concluded: "You worry about the so-called "persecution" of a few intellectuals. But these are an insignificant minority. We do not care about them and we will not let them destroy our system. We care about the large majority of our people". Then he accused us of putting pressure on them about Shklovsky and other dissidents. He said that the Soviet Union was big enough, not to tolerate any pressure. "If you continue putting such a pressure on us, you are going to lose" he said.

As he was in such a candid mood, he started speaking about the leading Soviet astronomers. He attacked particularly Zeldovich, who had an independent power of his own, outside the bureaucracy represented by Khromow. I told him: "This is your mistake. You cannot afford losing people like Zeldovich, Shklovsky or Zakharov. These are among the most brilliant Soviet scientists. It is because of such people that Soviet science is so famous today. You cannot reduce them to simple executives of your system".

Despite our disagreement we remained good friends. And Khromov would send me every Christmas a card with the picture of a nice russian orthodox church. In 1990 he wrote me "this time we are celebrating Christmas too".

When I finished my term as General Secretary of the IAU, at the Grenoble General Assembly (1966), both Massevich and Khromov came to thank me for what I had done for the Soviet astronomers. I was happy. They realized that I was quite sincere when I was saying that I cared about all astronomers, and the Soviet astronomers with all their difficulties were foremost in my mind.

22. TOPOLOGICAL METHODS IN GALACTIC DYNAMICS

During my stay in Columbia University, Harvard University and MIT in 1968-1969, I worked on a topic that puzzled me for a long time. What happens to dynamical systems when the perturbations are large instead of small? The theory of small perturbations was well developed at that time. One could apply perturbation theory to derive the main characteristics of the dynamical systems. This perturbation theory included the use of third integrals, and the Kolmogorov-Arnold-Moser (KAM) theory. In this way we could find analytically the forms of the orbits (Fig. 21a), the positions of the periodic orbits, and the statistics of the orbits of stars in a galaxy, that were then verified by numerical integrations.

But in the case of large perturbations the analytical methods are not applicable in general. In these cases most orbits are chaotic (Fig. 21b). Then our main tool is based on numerical explorations. Of course these numerical explorations are guided by continuity theorems, like those of Poincaré on the existence of families of periodic orbits, on homoclinic and heteroclinic orbits, etc. Local

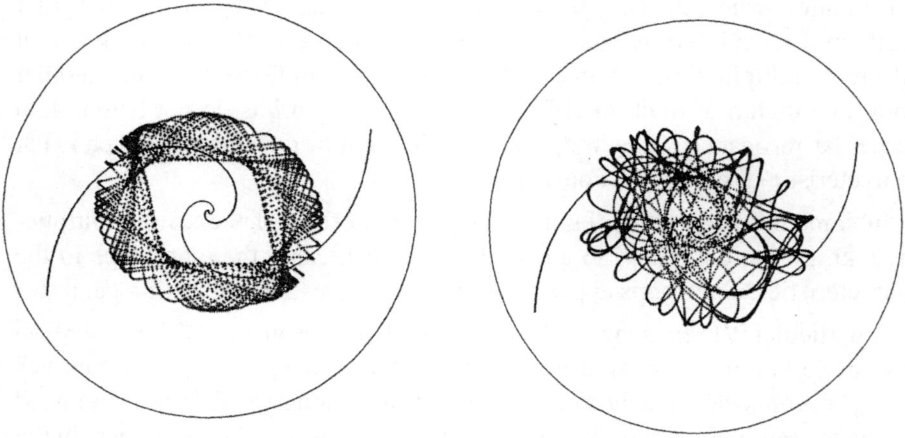

Figure 21. An ordered orbit (a), and a chaotic orbit (b) in a spiral galaxy.

theorems can always provide useful information. But the global structure of dynamical systems has to be explored numerically.

I did a systematic study that resulted in three papers in the Astronomical Journal on "Highly Perturbed Dynamical Systems":

(1)Periodic Orbits (Contopoulos 1970b)

(2)Stability of Periodic Orbits (Contopoulos 1970c) and

(3)Nonperiodic Orbits (Contopoulos 1971b).

The first paper contains a study of the periodic orbits and of their bifurcations. As the energy increases, the dimensions of the orbits change continuously. The size of an orbit as a function of its energy is called a "characteristic". At certain values of the energy new families are generated from the original family. These are "bifurcating families" with periods equal to multiples of the original period. But besides the usual bifurcations from the families of the unperturbed problem we found many new families of "irregular" orbits, i.e. orbits not connected to any of the families of the unperturbed problem.

The regular periodic orbits have either a single intersection with a Poincaré surface of section or a finite number of intersections, e.g. 3 or 5 intersections. In the latter case they have well defined "rotation numbers", i.e. their successive intersections by the Poincaré surface of section rotate with a well defined average rational angle $\theta/2\pi = n/m$, e.g. 1/3 or 1/5. The bifurcating families are of multiplicities 3, 6, 9..., or 5, 10, 15..., but their rotation numbers are again 1/3, 1/5. Around every point of a triple periodic orbit there may be two points of multiplicity 6, three points of multiplicity 9, etc. But the irregular families are very different. They are generated at a "tangent bifurcation", i.e. a stable and an unstable family are both generated at a point with a minimum (or maximum) perturbation, where the tangent of their characteristics is vertical. The irregular orbits may have intersections with a Poincaré surface of section in regions of different multiplicities, e.g. one point in a region of multiplicity 3 and another point in a region of multiplicity 5. Therefore, these orbits do not belong to a particular resonance. They indicate an interaction of resonances, which is the characteristic of chaos (section 13).

In some cases one finds families of periodic orbits of the same multiplicity, that are not connected to each other. In such cases there are gaps in the characteristics. Such gaps appear near the resonances 2/1,4/1,... of a galaxy.

A particular 2/1 resonance is the Inner Lindblad Resonance (ILR), where we have three families of oval orbits, known as families x_1, x_2, x_3. The families x_2, x_3 are connected, but the family x_1 is separated by a gap. This gap is reduced when the amplitude of the spiral perturbation decreases and it disappears in the limiting unperturbed (axisymmetric) case.

These examples led to the suspicion that all irregular families could be joined together if one is allowed to vary not only the energy, but other parameters also (in the galactic case a second parameter is the amplitude of the spiral). But we could prove that in most cases this is not possible.

Namely, we proved that "genuine" irregular orbits appear in regions of chaos, i.e. inside the lobes of the homoclinic tangle of unstable periodic orbits (section 43). Such orbits cannot be joined by a continuous characteristic with the regular families that appear outside the lobes, because such a curve would cross a side of the lobe, whose points do not represent periodic orbits but asymptotic orbits. This is a vivid example of a topological theorem that cannot probably be derived by other analytic methods.

I was speaking at a meeting about this subject when one participant interrupted me, claiming that he would be able to join all irregular families if I would allow him to use further additional parameters. I tried to answer him but he insisted. He even said "Give me the initial conditions of your orbits and I will show you how they are joined". I asked him "How long do you need to do that?". "One month". Then I said: "Let us have a bet. If you do not find the connections within a year, I will have won". I gave him all the data (and the dynamical model, which was extremely simple), but he never found the connections that he claimed.

My second paper, on the stability of periodic orbits, studied the changes of stability that appear at various bifurcations. It is important to note that the stability type of an orbit (stable vs unstable) can change only at particular types of bifurcations. This paper contains a lot of analytical results, including a complete classification of the resonant bifurcations. However, all these results are local. They do not provide information about the global structure of the bifurcations for large perturbations. We could only state that between the bifurcations of families with rotation numbers n/m and n'/m' there are bifurcations with all intermediate rationals.

On the other hand the numerical evidence indicated that beyond a period doubling bifurcation there are infinite bifurcations by successive period doublings (orbits of period 4, 8, 16, etc). At every bifurcation the original family becomes unstable, but its stability is inherited by the new, double period, family.

I noticed that the intervals between successive bifurcations decreased. But I did not calculate the rate of decrease. It was found later by Feigenbaum (1978) and independently by Coullet and Tresser (1978) that the decrease is almost geometrical (section 29).

Another aspect of this problem was prominent at that time, whether there are always stable periodic orbits for arbitrarily large perturbations. I had a correspondence about that subject with Joe Ford from Atlanta (Georgia Institute of Technology). We agreed that there were probably infinite period doubling bifurcations within a finite interval of values of the perturbation parameter ε. Beyond a limiting value of $\varepsilon = \varepsilon_\infty$ there should be infinite unstable periodic orbits. However I noticed (Contopoulos 1975b) that beyond ε_∞ there were values of ε with new, irregular, families of periodic orbits. All these families started as a pair of a stable and an unstable orbit. The stable family became unstable at a period doubling bifurcation, and for even larger ε there were infinite period doublings, and a new infinity of unstable families.

The question now was whether such stable families appear for arbitrarily large perturbations. In some cases we could prove analytically that this is the case. Namely some families have infinite transitions from stability to instability and back to stability for arbitrarily large ε. Therefore there is no critical value of ε beyond which only unstable families appear. This has important consequences as regards the ergodicity of a system. If there are islands of stability the system is not ergodic.

The appearance of islands of stability for arbitrarily large perturbations has been proven recently (numerically and analytically) for generic dynamical systems like the standard map (Dvorak et al. 2003). What remains to be seen, is whether there is genuine ergodic behaviour between values of ε containing islands of stability. This problem is still under study.

The third paper of the series dealt with nonperiodic orbits. Some orbits are ordered, forming closed invariant curves on a surface of section, while other orbits are chaotic, filling a domain of the surface of section with scattered points.

An important question, discussed in that paper, is whether the chaotic orbits in the same chaotic domain, are in some way equivalent or not. It was noticed that some initial conditions lead to different distributions of the successive points of an orbit in the chaotic domain. In particular, orbits starting close to an island of stability remain close to this island for a long time, before reaching the outer chaotic domain. This phenomenon was the first example of what was called later "stickiness", and it is described in more detail in section 47 (Fig. 35).

I found this phenomenon by trying to find numerically the last KAM curves that limit islands of stability. The orbits inside the islands remained there for ever, while the orbits a little further away stayed there for a long time, but then they escaped to the large "chaotic sea". I thought, first, that this was a numerical effect. Thus, I improved the accuracy but the effect was the same. Finally I made the following check. After the orbit entered the chaotic domain I reversed the velocity and continued the calculations. Then I saw that the new orbit followed the previous steps backwards and finally it was trapped near the boundary of the islands. Thus, the effect of stickiness was real.

This topological effect appears in many chaotic domains of dynamical systems.

I presented the topological methods of stellar dynamics at the 1973 winter school in Saas-Fee, Switzerland, on the "Dynamical structure and evolution of stellar systems" (Contopoulos 1973a). This was one of the best schools that I attended. There were only three speakers, Donald Lynden-Bell, Michel Hénon and myself. The classes were taking place in the morning and in the evening, while the afternoon was devoted to skiing. I did not ski, but I had some long walks high up in the mountains, that were covered deeply by snow.

Lynden-Bell spoke about resonant orbits in galaxies, the fragmentation theory and the statistical mechanics of stellar systems. Hénon spoke about the collisional dynamics of stellar systems. I spoke about the density wave theory and the use of topological methods in stellar dynamics. The topological methods are especially useful for systems with large perturbations.

Lynden-Bell (1998) wrote about this meeting: "In 1973 at Saas-Fee, George gave lectures in which he introduced me to the wonders of modern dynamical theory- topological models, incomplete chaos, and the KAM theorem. It opened my eyes to so much that was new to me!".

23. THE IAU GENERAL ASSEMBLY IN GRENOBLE

My last job as General Secretary of the IAU was the organization of the Grenoble General Assembly in 1976.

The local organizers were extremely helpful and effective. But I had to supervise the scientific programs of the Assembly and that was extremely complicated. There were meetings of about 50 Commissions, several invited lectures, joint discussions, meetings of the national representatives of about 50 countries, of the presidents of the Commissions, of the Executive Committee, of the Finance Committee, etc. We had also the two main sessions of the General Assembly. In total there were about 250 separate meetings, and I had to write an innumerable number of letters to arrange everything in a proper way.

Then there were problems of every kind. I give only two examples here.

(1) Shklovsky was not allowed to come as an invited speaker and had to be replaced the very last moment.

(2) Carl Sagan, who was an invited speaker was supposed to speak about the latest results of the Mars missions. He sent me a letter asking to cover his trip by Concorde in order to bring the very last results from the Viking space probes. I sent him a regular air ticket without any comment.

I had arranged to spend the year after the General Assembly at the European Southern Observatory (ESO), which had its headquarters inside CERN in Geneva. I had rented a house in Prevessin, France, just across the border from the entrance of CERN, in Switzerland.

Thus, I took my family in my car and travelled by ship to Brindisi, and then by car all the way to Geneva. We stopped on the way in Rome, Florence and near Milan. But the children suffered from such a long tight confinement in the car (4 children) and they were continuously complaining. After some time I had a good idea. I started telling them fairy tales. The children were then so absorbed by these stories that they forgot to complain.

We reached our house near Geneva and then moved to Grenoble. In Grenoble I was completely absorbed by my duties. I told my family that they had to be patient, and I added: "If I survive this General Assembly, I will live very long".

I had no rest during the General Assembly, but things went more or less smoothly and we reached the closing session of the General Assembly. During this session the President of the IAU, Leo Goldberg, said: "During the last three years, I have watched with wonder and admiration as George Contopoulos, with the loyal and hard-working assistance of the executive secretary Arnost Japperl and the administrative assistant Jarka Dankova, have handled literally thousands of pieces of correspondence, administered the funds of the Union, prepared and issued the publications and worked out the intricate details of the program of the scientific and administrative meetings during the General Assembly. What is most remarkable about your performance, George, is that you have been able to continue your outstanding research in galactic structure, as those of us who have heard your lecture recently can testify". This last remark was true. In fact I worked hard during my term as General Secretary to continue my research, and I was happy to see that this was acknowledged. At the end Dr. Goldberg said: "Finally, I want George Contopoulos to know how much I appreciated the tact and diplomacy with which he educated me to the responsibilities and especially to the prerogatives of our respective offices". It was clear that he realized that I had to overcome many difficulties in order to bring about a harmonious collaboration with him for the benefit of the IAU.

After that it was my turn to speak. I said what I felt most strongly at that time. "Perhaps I am the most happy person of this General Assembly at this moment, as my task is over". I thanked all my collaborators and concluded: "I would like to finish with a small personal story. As I was absorbed by the work of the IAU together with my other duties, research, administration and travel, my children complained that they did not see enough of their father. And during one of my travels, my little daughter improvised a new kind of prayer: "Let daddy come back to us, and not be lost in some foreign airport". Well, thank God, I was not lost anywhere and I am happy to be here with you today. But I am also happy to return to my family and to my research.".

The end of the General Assembly (Fig. 14) was, in fact, the most happy moment, after so much work and efforts over the previous years, and, in particular, the previous weeks. Everyone in the IAU seemed happy, and I felt a free man again.

But that session was not quite the end. After that I was asked to attend the first meeting of the new Executive Committee as a consultant. I had only to give some information and some suggestions. But the meeting lasted several hours. Meanwhile my wife expected me in the hotel and she was very concerned, because I had told her that I should return soon. She had been patient with me for six whole years, as I was absorbed with the IAU, but these extra hours were too much for her. At any rate she was greatly relieved when she saw me. We moved to the car and drove all the way to Geneva. A new life started for us at that moment.

24. A SABBATICAL AT ESO

I spent the year 1976-1977 at the ESO headquarters in Geneva. It was a very fruitful period, but with many difficulties. Every day I crossed the frontier, going from the village Prevessin in France, to CERN in Switzerland. I usually went by bicycle, under all weather conditions.

My children went to schools in France. But we were disappointed with the local school in Prevessin, where the children spent the day without learning anything essential. No homework was assigned to them, and it was forbidden to do any reading at home. But fortunately we found an excellent private school in Switzerland, in Versois. It was also fortunate that there was a school bus every day from Prevessin, France, to Versois, Switzerland. Thus, my three girls went to this school. The two older girls could manage enough French to attend the school, but my wife spent a lot of time every day to help them. I also helped them in specialized topics like mathematics when I was returning home late in the evening. The problem was with my youngest girl, Emmanuela, who did not speak any French. Because of that she was very unhappy at first. But she soon made friends there and somehow she could communicate with them. She had never been in a school before, and as she had one more year to enter the primary school in Greece, I thought that she attended a kindergarten class. Only at the end of the year I realized that she had finished the regular first class of the primary school.

At first she refused to try to speak French in class, but one day she burst out with a torrent of French words. I learned that she could also write, in a very unexpected way. One day I got a telegram that my father had died, and I rushed to take a plane to go to Greece for his funeral. As I was leaving, Emmanuela gave me a piece of paper. She said: "This is a letter for grandma". I tried to read it and suddenly I realized that it was Greek with French characters. It said: "Dear grandma don't cry. Grandpa is in heaven".

The end of the story was at the closing ceremony of the school in Versois. We were happy to hear that our oldest daughter got the third prize and our second daughter the second prize, of their respective classes. But the real surprise was our youngest daughter, who got the first prize of her class. It was nice, when all three went up, dressed in very similar pink dresses, to get their diplomas (Fig. 22).

On the other hand our son (our oldest child) went to an international school in a nearby town called Ferney-Voltaire. He attended a section of the school that had English as second language. His teacher asked me to visit her in the school and said: "Why do you send your child to the English section? He does not speak English". I said: "Please have patience for one month. If he does not speak English then, send him away". But my son managed to learn enough English in one month, to be accepted in his class. Then he did something more daring. As I was leaving for a series of lectures in England he told me: "Dad. The level of our class is too easy for me. So I want to go to the next higher class". I told him not to say any such nonsense, and I left. When I returned from England, a few days later, my son told me that he had seen the Director of the school, asking to be transferred to the higher class, and the Director accepted his transfer after consulting his teachers.

During my stay in ESO, I did a lot of research work with the ESO computers. I had followed the development of the computers since my early calculations in Stockholm in 1956. At first we were using punched cards, later various typewriters and then windows. I was spending innumerable hours in front of the windows of the CERN computers, that were much faster than any computer that I had used so far.

Whenever I did not type any instruction for my computer for a long time a message would appear in front of me "Are you bored?" "Yes" I typed. Then the computer started to print some jokes. It is in this way that I learned the various versions of Murphy's law. Examples: "If anything can go wrong, it will go wrong", and "when one corrects an error in a computer program, he makes on the average three new errors". I was surprized because I had devised independently a similar "law" myself. I was saying that one of the dwellings of the devil is the computer. The devil does all kinds of tricks to make us fail in our efforts.

Figure 22 My daughters after the Versois (Geneva) school ceremony (1977).

Figure 23. At the LaPlata meeting (1995) with J. Henrard.

During our stay in ESO we had many opportunities to go around with my family. We went to Chamonix, Jura, Zermatt and other winter resorts, and to many places in Switzerland, France, Germany and Italy. We used to go to a Greek orthodox church in Switzerland. I remember the Easter night when we were returning to our house in France with our Easter candles. The frontier guards were quite surprised to see us with so many candles lighted.

One of our longest trips was to Laco di Como in Italy. There was a meeting on "Stochastic behaviour in classical and quantum Hamiltonian systems" organized by G. Casati and J. Ford. The meeting was very interesting, and I had also the opportunity to take my family around Northern Italy and Southern Switzerland.

The summer of 1977 I took my children to a camp south of Grenoble, organized by Russian immigrants. It was a very good camp with many activities, including music. The Russians organize very nice choirs and we enjoyed considerably some evenings that we spent there, when we visited our children.

Finally in August 1977 we started our long trip home. We drove all the way from Geneva, Mt. Blanc, Northern Italy to Venice. Then after a tour of the city we took the boat to Greece. We came home very tired but happy.

My children went to four different schools in Greece. The schools were good, but their schedule was quite complicated. In some cases there were no school busses. My wife spent a lot of time, every day, to take some children to their schools, to their music lessons, etc. But the children adapted well to their new environment. In particular my youngest daughter was elected president of her class. She had told me "I want to become president. Can you give me money to buy a candy for all the children to vote for me"? I gave her the money but she preferred to keep it for herself, and she only did a persistent lobbying, that was successful. I congratulated her for her success, adding: "Now you are a comrade president". At that she felt slightly indignant. "What kind of president are you"? she said.

25. ASTRONOMY AND ASTROPHYSICS

One of the main European activities of ESO was the European Journal "Astronomy and Astrophysics". This journal was formed by the merging of the main national astronomical journals, namely the French journals "Annales d'Astrophysique", "Bulletin Astronomique", and "Journal des Observateurs", the German "Zeitschrift für Astrophysik", the Dutch "Bulletin of the Astronomical Institutes of the Netherlands", and the Scandinavian "Arkiv för Astronomi". This was materialized at the Prague General Assembly of the IAU, in 1967, and the first president of the new journal was Dr. A. Blaauw, then Director General of ESO. The European Southern Observatory provided financial facilities for the journal, including a secretariat for the Board of Directors.

The merging of the major European journals was a major achievement. Instead of being antagonistic these journals joined forces in forming a robust European journal that became later the second largest astronomical journal in the world, after the Astrophysical Journal.

At first the British representatives had agreed to join Astronomy and Astrophysics. However the Royal Astronomical Society decided to continue the Monthly Notices, therefore the British remained outside the journal. Later on the Czech journal "Bulletin of the Astronomical Institutes of Czechoslovakia" merged with "Astronomy and Astrophysics".

During the Prague meeting of the IAU in 1967, Dr. Z. Kopal announced the launching of a new private journal under the title "Astrophysics and Space Science". He had a list of leading scientists that were members of the Editorial Board. In Prague he asked me to participate in this Editorial Board, but I declined. I told him that at a time when a reduction of the number of journals was achieved by forming the European Journal "Astronomy and Astrophysics" I did not think that a new journal should be formed as an opponent. Then Kopal told me that if I joined his Editorial Board, he would show me a paper scheduled for the first issue of the new journal, which attacked my "third integral". I said that if such a paper appeared I supposed that I should be allowed to answer. "Yes, he said, but it would be better for you to have your reply in the same issue". (This is done as a routine in journals like Astronomy and Astrophysics. It is never considered as a favour). "You do what you like" I replied. Thus, this paper appeared in Astrophysics and Space Science, in 1967. It was claiming that the third integral is a function of the Hamiltonian. My reply appeared later as a joint paper by Contopoulos and Barbanis (1968) under the title "Is the third integral a function of the Hamiltonian?"[7].

The journal of Dr. Kopal was run without systematic refereeing. I was told that in many cases the refereeing was done by Dr. Kopal himself, or by another editor.

---

[7] The simplest answer in the case of a Hamiltonian

$$H = \frac{1}{2}(\dot{x}^2 + \omega_1^2 x^2) + \frac{1}{2}(\dot{y}^2 + \omega_2^2 y^2) + \varepsilon H_1 + ... \qquad (25.1)$$

is that if the third integral

$$\Phi = \frac{1}{2}(\dot{x}^2 + \omega_1^2 x^2) + \varepsilon \Phi_1 + ... \qquad (25.2)$$

is a function of H then for $\varepsilon = 0$ the integral

$$\Phi_0 = \frac{1}{2}(\dot{x}^2 + \omega_1^2 x^2) \qquad (25.3)$$

should be a function of the zero order Hamiltonian

$$H_0 = \frac{1}{2}(\dot{x}^2 + \omega_1^2 x^2) + \frac{1}{2}(\dot{y}^2 + \omega_2^2 y^2) \qquad (25.4)$$

which is impossible.

On one occasion I asked Chandrasekhar, whose name appeared in the Editorial Board why he had accepted to be in this list, and he said: "It is good to have a second class journal, to absorb the burden of mediocre papers".

The European journal "Astronomy and Astrophysics" had a spectacular development in the coming years.

Some years later I was elected president of the Board of Directors of "Astronomy and Astrophysics" (1979) and I was re-elected every three years until 1993, when I stepped out.

Many improvements were done in the Journal during that period. The size of the journal increased considerably, from 9 volumes of 400 pages in 1979 to 33 volumes in 1993. Thus, Astronomy and Astrophysics went above the Monthly Notices and the Astronomical Journal as regards the number of pages and only the Astrophysical Journal is now above it.

The number of countries belonging to ESO was increased from 10 to 17. The new countries were Spain, Finland, Check Republic, Slovakia, Hungary, Poland and Estonia.

The journal had two publishers. The main publisher was "Springer Verlag" from Germany, who published the Main Journal, while the Supplements were published by the French "Editions de Physique".

There were many negotiations with the publishers aiming at reducing the price of the journal. The negotiations were tough. On one occasion, when we discussed the bids of various companies that wanted to take over the journal, we had an offer by Les Editions de Physique that was much better than the offer of Springer Verlag. Thus, the Board decided to give the Main Journal to Les Editions de Physique. But then Springer Verlag came with a lower offer, equal to that of Les Editions de Physique, and we decided to continue with Springer Verlag, whose competence was tested during so many years.

The financial situation of the Board of Directors improved considerably. When I took over the Chairmanship, we were in the red and we survived only due to the generosity of ESO. When I left the journal we were in a very safe situation, despite the fact that on many occasions we had to subsidize the journal, especially in publishing extra volumes (without extra charge for the subscribers), or in distributing free copies to Eastern Countries, and Institutes that had financial difficulties.

When I left the Journal the new Chairman, Dr. A. Maeder, from Switzerland, wrote me: "Please allow me to express my gratitude both personally and in the name of the Board of Directors of Astronomy and Astrophysics for all the actions you have undertaken in favour of the European Astronomy as Chairman of this Board. During 14 years you have worked for the success of the European Journal, defending the quality, the reputation, the management and the financial health of the Journal, which has now become, as a result of your efforts, the second major journal in the world. You have conducted the Board with great efficiency and authority, always maintaining a good and warm atmosphere during our meetings. The astronomical community is greatly indebted to you. As your successor I can take over Astronomy and Astrophysics in very good shape and the work will be continued in the same spirit you have shown us".

The Editors of the Journal did an excellent job and improved the quality of the papers, by using a strict refereeing system. Among the Editors I should mention Drs. Lequeux, Cesarsky and Praderie from France, Dr. Grewing from Germany, and the letter Editor, Dr. Pottasch from the Netherlands. I had an excellent collaboration with all of them, but I never interfered with their job. In particular my own papers were refereed in the most strict way. I received a few complaints from some authors, but I referred them to the Editors. The same attitude was followed by the Board of Directors.

At one meeting of the Board we discussed the citations of the main astronomical journals. It was noted that while there are many citations from Astronomy and Astrophysics to Monthly Notices and vice-versa, there was not a similar reciprocity with the American journals. E.g. the authors of Astronomy and Astrophysics cite very often papers in the Astrophysical Journal, but many authors of the Astrophysical Journal ignore related articles in European journals.

It seems that some people are limited within their own horizon and do not care to look beyond this horizon. E.g. some Americans ignored important results that appeared in Europe, including the Soviet Union, or in Japan. I should mention a notorious example, the discovery of the universal sequence of bifurcations leading to chaos (section 29). This was found independently by Feigenbaum in the United States and by Coullet and Tresser in France, both in 1978. In particular the intervals between successive period doubling bifurcations decrease almost geometrically, with a universal ratio $\delta = 4.67$. The work of Feigenbaum was cited several times by Coullet and Tresser, but Feigenbaum never referred to Coullet and Tresser.

Of course the tendency to overemphasize one's own work and ignore the work of others is very common. Another well known example is the discovery of inflation in Cosmology. This was established in a paper by D. Kazanas in the Astrophysical Journal Letters, in 1980. Kazanas (1980) called this phase of the early Universe "de Sitter's phase". The name "inflation" was given by Guth (1981). But Kazanas had already given the basic elements of inflation, the inflation era starting at the GUT (Grand Unified Theory) time $t = 10^{-35}$ sec, the supercooling period, the reheating of the Universe, followed by an exponential expansion, the transition to normal expansion, etc[8].

The paper of Kazanas attracted little attention for a long time, while Guth, who published his results in the Physical Review one year later (1981) became

[8]A few elements of the inflationary period had been already discussed by some Russians, like Starobinski.

famous. Only later the work of Kazanas and of others was appreciated. E.g. the journal Mercury published in April 1989 a picture of the abstract of the early paper of Kazanas with the remark that Kazanas preceded Guth in the theory of the inflationary universe. Guth himself did not refer to the work of Kazanas, until he published a book on the subject under the title "The inflationary universe; the quest for a new theory of cosmic origin" (1997), where he apologizes for not having referred to the early work of Kazanas and of others, related to inflation.

26. TRAVELLING IN WESTERN EUROPE

Starting in 1955 I have travelled a lot all over the world, but mainly in Europe and the United States. My first trips in Europe were by train, or bus, but later I used almost exclusively the plane. Most of my trips were rather long and with several planes in succession.

Only on one occasion I could return to Athens the same day from abroad. In 1978 I had a meeting in Rome, and I flew there early in the morning. I was supposed to stay there one night, but the meeting ended at 2 p.m. Thus, I went to the airport and I succeeded to change my ticket for the same day. When I arrived in Athens and came in front of our home I called my wife. I wanted to make a joke and told her that the meeting was extended for one more day. Then two minutes later I opened the door of my house. You can imagine the surprise of everyone.

Many trips were related to my positions in the IAU, the Board of Directors of "Astronomy and Astrophysics", the European Science Foundation and the Science Committee of NATO. I had also many invitations to scientific meetings, or to give seminars at various Institutions.

In my diary I see my trips in the fall of 1977, after my return to Greece from ESO in August 1977. End of August: Geneva (Executive Committee of the IAU); September: Warsaw, Torun (IAU Colloquium No 65), Geneva; October: Geneva, Brussels (NATO Science Committee), England (lectures at Queen Mary College, Cambridge and Oxford); November: Geneva; December: Geneva, Amsterdam, Geneva. And so on.

My trips were not always uneventful. During a meeting in Italy I had a serious problem. The meeting was in the island of Capri (1990). I arrived in Rome and I should take a train to Naples and then a boat to Capri. I bought the train ticket, but when I boarded the train I noticed that my wallet with all my money was stolen, during the congestion of people boarding the train. I went to the police, but they could not offer me anything, besides writing a report. Thus, I took the train and arrived in Naples. But the boat would not take me to Capri without a ticket. Fortunately the offices of the boat company were not far away. There I explained my problem and they found a solution. They called the hotel where we had the meeting in Capri and I could speak with one of the organizers, Dr. A. Giorgilli. He arranged with the company that he should be at the harbour of Capri when I should arrive and pay my ticket. Later he provided me with enough money for my local expenses and for my trip back.

On another occasion I had an accident in my hotel in London, I slipped in my bath and I had a deep wound in my head. I was covered with blood. I tried to stop the blood and called the ambulance. They took me to a nearby hospital and put several stitches. Fortunately I could continue my trip the next day to Cambridge.

Whenever I visited Cambridge I stayed in the home of Donald Lynden-Bell. The first time I went there I was supposed to call him from the train station. In order to speak to him I should press a button after hearing his voice, but, instead, I spoke to him without pressing the button, and I was not heard. I tried again, and this time I heard Donald, who guessed who was trying to call, saying: "George, press the button".

During my visits to Cambridge I was often walking with Donald Lynden-Bell from his office to his home. He would take this opportunity to discuss many scientific problems with me, ideas that were not yet clearly formulated, and prospects for future work.

From Cambridge, I used to take the bus during a weekend to a Greek orthodox monastery in Tolleshunt Knights, Essex. This is a very remarkable monastery. It was established by a Russian monk, father Sofrony Sakharov, and it consisted in fact of two nearby monasteries, one for monks and the other for nuns.

The monks and nuns were from many countries, Russia, England, France, Germany, Belgium, etc. Only two persons were Greek. One of them was a Greek from Australia; he had graduated in chemistry and then he got his PhD in London. There, he met father Sofrony and he decided to become a monk. Later he succeeded father Sofrony, as abbot.

Father Sofrony had a remarkable history. He was a young painter in St. Petersburg during the Revolution. Then he emigrated to Paris and he was a quite successful painter there. But his search for a deeper meaning of life brought him to Mount Athos in Greece, where he met a Russian monk, father Silouan, a very remarkable spiritual man, who was later canonized. After many years he left Mount Athos and came to Paris, to publish the biography and the teachings of St. Silouan. There he fell very seriously ill. He was hospitalized and they expected him to die. Instead, he had a very remarkable recovery and many people approached him after his book was published. This book had a large number of editions in many languages.

He decided to go to England, to Tolleshunt Knights, and he established a monastery in an old farm close to an old gothic church. Later this monastery was made known all over the world and many people were coming there as visitors or as permanent members, attracted by the saintly personality of its abbot. Father Sofrony was a remarkable mixture of a deeply spiritual person and an intellectual of very high caliber. He spoke several languages, Russian, English, French, Greek, etc. I had many discussions with him and I always remember these discussions. He might start in Greek, but then he found more convenient to speak in French, or English.

The services in the church were also in many languages, depending on the audience. Part in Russian, part in English, Greek, etc. My wife and children were very impressed by him. During the difficult years as late teen-agers my children would often spend their vacations there.

The monastery was very poor, but it always provided the usual hospitality of Greek orthodox monasteries. People would stay there and take part in the frugal meals of the monks.

I remember when I first went there. I arrived in Chelmsford and took a taxi. The taxi driver had never heard of Tolleshunt Knights and we spent a lot of time in the plains of Essex trying to locate it. When I arrived the monks were very friendly. The spiritual atmosphere was excellent.

Later, this monastery was well known by the local people and by visitors from all over the world. On one occasion I was at the train station trying desperately to find a taxi, when a local person approached me and asked: "Are you going to the monastery of Tolleshunt Knights"? "Yes" I said. And he volunteered to drive me there, free of charge, although he was going to a different town.

In 1990 I was in London with a serious health problem. I had a tumour and the doctors in Athens had suspected cancer. They did not tell me that, but they informed my wife, who is a medical doctor. We went to a hospital in London for a check-up, and then we visited the monastery in Essex. Father Sofrony asked me to kneel down and recited an impressive long prayer. After that we returned to London to learn the results of the biopsy. The doctor called the lab, and they told him that my tumour was benign. We were greatly relieved. But my wife was so excited that, when she heard the word "benign", she fainted and fell down. It took the doctor several minutes to restore her. And only then she told me that the doctors in Greece had suspected cancer.

Father Sofrony died in 1993, but his spirit still lives there. There are now many publications containing the works of father Sofrony. And many people visit continuously the monastery to find solace and spiritual guidance.

27. DESTRUCTION OF THE INTEGRALS

During my stay in ESO and after my return to Greece I was interested in the problem of the destruction of the third integral, following my earlier work on the overlapping of resonances (section 13). This phenomenon refers to two or more degrees of freedom. In two degrees of freedom the "third" integral is, in fact, a second formal integral.

A particular case of resonance interaction in systems of three or more degrees of freedom is when the unperturbed frequencies satisfy more than one resonance conditions. It has been known for a long time that in such a case the number of formal integrals is reduced. An example was given in action-angle variables by Ford (1975):

$$H = \omega_1 I_1 + \omega_2 I_2 + \omega_3 I_3 + \varepsilon[\kappa \cos(2\theta_1 - \theta_2) + \lambda \cos(\theta_1 + \theta_2 - \theta_3)] \quad (27.1)$$

where $\omega_1 = 1, \omega_2 = 2, \omega_3 = 3$ and κ, λ are functions of I_1, I_2, I_3.

In this case there are two resonances, $\omega_2 = 2\omega_1$ and $\omega_3 = \omega_1 + \omega_2$. It can be easily seen that $\Phi = I_1 + 2I_2 + 3I_3$ is an integral of motion but there is no further integral.

Ford noticed that the system $H' = H - \Phi$ is chaotic for any value of ε. In fact a change of ε does not change the phase space of the system but only the time scale of describing its orbits. Ford then called this case "a new type of instability" because this instability is introduced for arbitrarily small values of ε.

But I did not agree with this characterization and I studied what happens when the resonances are not exactly fulfilled (Contopoulos 1978b). I noticed that another parameter, besides ε, is the deviation from the exact resonance, e.g. $\omega_2' = \omega_2 - 2\omega_1$. If ω_2' is not very small and ε is small with respect to ω_2' then there are three integrals of motion. If, on the other hand ω_2' is very small and ε not so small the correct parameter to use is $\varepsilon' = \varepsilon/\omega_2'$ and this may be large.

Therefore the correct expression is that an integral is destroyed not for arbitrarily small ε, but for $\varepsilon' = \varepsilon/\omega_2'$ larger than a critical value.

The form of the integrals of motion in a particular resonance is different from its form far from resonances. An interesting problem was to find the change of the integrals as we approach a particular resonance. Near the resonance the forms of the orbits change and new types of resonant orbits are introduced (Contopoulos 1965a, 1968, 1976)(Fig.4).

The use of appropriate variables and of a sufficiently large number of terms of the third integrals is important in finding sufficiently accurate results in the case of ordered (nonchaotic) orbits. E.g. in applying the theory of the third integral to the Trojan asteroids around the Lagrangian points L_4 and L_5 in the solar system it has been necessary to use appropriate variables and expansions up to order 30 (Skokos, Contopoulos and Giorgilli 1996; Giorgilli and Skokos 1997). But when the orbits become chaotic the theory cannot be applied. Then chaos is introduced, and we say that the third integral has been destroyed.

This can be seen in a spectacular way in the case of a spiral galaxy. If the spiral field is weak the orbits are mostly ordered (Fig.14a), filling elliptical rings around the center. But if the spiral field is stronger most orbits are chaotic (Fig.14b).

28. SYSTEMS OF THREE DEGREES OF FREEDOM

For many years I have been interested in dynamical systems of three degrees of freedom. One of my first papers on this subject was done with Luigi Galgani and Antonio Giorgilli (Contopoulos et al. 1978), and dealt with the number of isolating integrals in Hamiltonian systems. In a time independent 3-D system we find in general ordered regions with three integrals of motion and chaotic regions with only one integral of motion (the energy). However, we were surprised to find also some small domains, where it seems that there are two and only two integrals of motion.

The phase space in 3-D systems is 6-dimensional, but, because of the integral of energy, it is effectively 5-dimensional. A Poincaré surface of section is 4-dimensional. It is not easy to visualize such a space, although we get an idea of its structure by using motion and colours. The third dimension can be seen, in principle, by moving our eye around an object. Instead of that we rotate a figure in the computer and get a good impression of the third dimension. The fourth dimension is represented by colours. Blue for points closer to us, red for points further away. Thus, two lines that intersect in three dimensions do not intersect in reality if they have different colours, because their 4th dimension is different.

A way to find whether an orbit is ordered or chaotic is by calculating its Lyapunov characteristic number. This is the average rate of exponential deviation of two nearby orbits. If the initial deviational time $t_0 = 0$ is ξ_0 and the deviation at time t is ξ, then the Lyapunov characteristic number is the limit

$$LCN = \lim_{t\to\infty,\xi_0\to 0} \frac{\ln|\xi/\xi_0|}{t} \tag{28.1}$$

In particular if the deviation is exponential

$$\xi = \xi_0 e^{at} \tag{28.2}$$

with $a > 0$, then the Lyapunov characteristic number

$$LCN = \frac{\ln e^{at}}{t} = a \tag{28.3}$$

is equal to the constant a. On the other hand if the deviation is linear

$$\xi = \xi_0 t \tag{28.4}$$

then the Lyapunov characteristic number is zero, because

$$LCN = \lim \frac{\ln t}{t} = 0 \tag{28.5}$$

In general ξ is given only numerically and LCN is found empirically. If $LCN = 0$ we say that the orbit is ordered and if it is $LCN > 0$ we say that it is chaotic.

The value of LCN in general does not depend on the direction of the initial deviation ξ_0. The LCN takes the maximum possible value for almost any initial condition. However, in systems of more than two degrees of freedom there are exceptional directions along which the LCN takes a different, smaller, value. This is a second Lyapunov characteristic number.

If a system has a second integral of motion, besides the energy, then the problem is reduced to a 2-dimensional one, and we have only one Lyapunov characteristic number different from zero. Therefore the existence of only one Lyapunov characteristic number different from zero indicates that there is a second integral of motion, but not a third one. The orbits are chaotic, but less chaotic than in the case of two LCNs different from zero.

In our numerical studies we found a "large chaotic sea" with two LCNs different from zero, and some "small chaotic seas" with only one LCN different from zero. In order to see the different chaotic seas we used stereoscopic pictures (Fig. 24). One has to see the left figure with the left eye and the right figure with the right eye. Some people manage to do that directly.

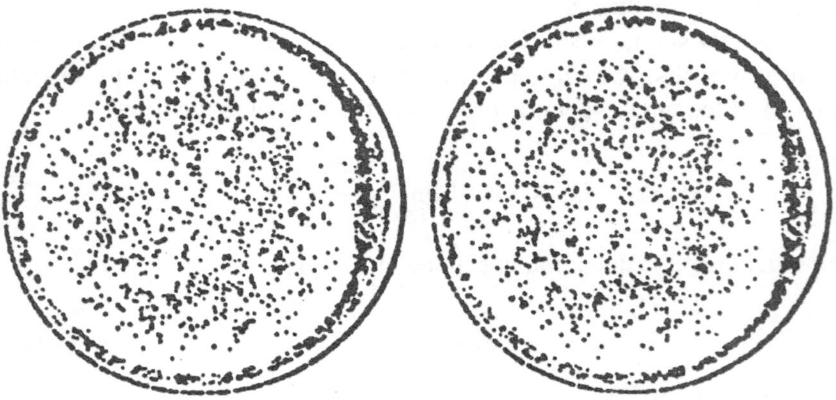

Figure 24. Stereoscopic pictures of a large chaotic sea (centre) and a small chaotic sea (periphery)

More surprising was the fact that the small seas did not communicate with the large sea over an extremely long time.

Theoretically it is expected that all chaotic domains in a system of three or more degrees of freedom communicate with each other. This phenomenon is called "Arnold diffusion" and it is due to the dimension of the phase space. Namely in a time independent system of N-degrees of freedom the phase space is of $2N - 1$ dimensions (because of the existence of the energy integral).

The regular orbits in such a system cover an N-dimensional surface (KAM theorem). If $N = 2$ then the N=2-dimensional surfaces separate the $2N-1 = 3D$ phase space. E.g. a toroidal 2-dimensional surface separates its interior from its exterior. But if $N = 3$ the N=3-dimensional surfaces do not separate the $2N-1$=5-dimensional phase space. Thus, orbits from one neighbourhood can travel to any other part of the phase space. This is Arnold diffusion.

In order to give a vivid example of this phenomenon let us reduce the number of dimensions by two. Then the phase space is 3-dimensional and the invariant surfaces are 1-dimensional, like strings from the floor to the ceiling. An orbit starting on a string cannot escape from this string.

If a system is integrable there are strings everywhere in the room and no diffusion is possible. However, if the system is nonintegrable there are everywhere empty regions between the strings. Therefore a careful person can avoid all the strings and move from one end of the room to the other.

But the time scale for Arnold diffusion may be extremely long. Thus, various chaotic domains may be practically separated, and if we stop our calculations after a large time (not infinite) we may find different Lyapunov characteristic numbers over a finite time. Our results opened a new interest in systems of three degrees of freedom. There are about 100 citations to our paper up to now, and many people extended our work with further examples (for more details see section 45).

An important problem in systems of three degrees of freedom is the study of periodic orbits. Our main papers on 3-D periodic orbits and their bifurcations are: Contopoulos and Magnenat (1985), Contopoulos and Barbanis (1985), and Contopoulos (1986a,b). In order to find the generic properties of the periodic orbits, we must consider at least four parameters. E.g. in systems representing three coupled oscillators

$$H = \frac{1}{2}(\dot{x}^2 + \dot{y}^2 + \dot{z}^2 + \omega_1^2 x^2 + \omega_2^2 y^2 + \omega_3^2 z^2) - \varepsilon x z^2 - \eta y z^2 \qquad (28.6)$$

we have two ratios of frequencies ω_2/ω_1 and ω_3/ω_1 and two parameters for the nonlinear perturbation, ε and η. Then we derive two stability parameters, b_1 and b_2 (Broucke 1969, Hadjidemetriou 1975).

We developed two types of diagrams for the various families of periodic orbits: (a) existence diagrams and (b) stability diagrams.

The existence diagrams give the domains of the parameter space (ε, η) (for fixed $\omega_1, \omega_2, \omega_3$) where a given family is stable or unstable, while the stability diagrams give the variation of the stability parameters as functions of ε, for particular fixed values of η, or as functions of η for particular fixed values of ε.

The periodic orbits are of four "stability types":

(1) Stable (when all eigenvalues are complex, on the unit circle). Then $|b_1| < 2$ and $|b_2| < 2$.

(2) Simply unstable (when one pair of eigenvalues is imaginary on the unit circle and the other pair is real). Then one stability parameter b_i is absolutely larger than 2, and the other is absolutely smaller than 2.

(3)Doubly unstable (when all eigenvalues are real). Then $|b_1| > 2$ and $|b_2| > 2$.

(4)Complex unstable (when all eigenvalues are complex but not on the unit circle). Then b_1 and b_2 are complex (Contopoulos and Magnenant 1985).

We adopted this simple classification, while Broucke (1969) used a more detailed classification.

The geometry of the diagrams follows certain rules that simplify their description. E.g. a stable orbit can become simply unstable, or complex unstable, but in general it cannot become doubly unstable. A simply unstable orbit can become stable, or doubly unstable, but not complex unstable, and a doubly unstable orbit can become simply unstable, or complex unstable, but not stable in general.

A more ambitious study refers to a change of the ratios ω_2/ω_1 and ω_3/ω_1 besides the changes of ε and η. I discussed the "qualitative changes in 3-D dynamical systems" in my 1986b paper, and found several patterns followed by such systems. In particular I could predict the main qualitative characteristics of a family of periodic orbits for any ratio ω_2/ω_1 by studying the problem only for a particular value of ω_2/ω_1.

I remember that the referee of my 1986b paper, while finding the paper interesting, complained that I had not given any forms of orbits and added: "I cannot see the forest because of the trees". In response I inserted a few characteristic types of orbits, but I remarked that the orbits correspond not to the trees, but to the leaves of the trees. Therefore if I would give further details about the leaves, the forest would be lost even more completely from sight.

A more detailed study of this problem, including a variation of both ω_2/ω_1 and ω_3/ω_1, was the subject of the thesis of an assistant of mine, L. Zachilas (1993). He prepared an "Atlas" of existence and stability diagrams that give detailed information about dynamical systems represented by Eq. (28.6).

Bifurcations of new families appear whenever we have transitions from one stability type to another (except type 4). Then a stability curve (giving a stability parameter as a function of ε or η) crosses one of the axes $b_i = \pm 2$. But we have further bifurcations when a stability curve reaches an axis $b_i = \pm 2$ without crossing it.

On the other hand a transition to complex instability does not produce a bifurcation in general. Because of that the scenario of infinite period doubling bifurcations, that leads to chaos (section 29) does not apply to 3-D systems. In such systems a sequence of bifurcations is terminated in most cases at a complex instability (or at an "inverse bifurcation"; see Contopoulos and Giorgilli 1988).

This phenomenon does not appear in systems of two degrees of freedom, where we do not have complex instability at all.

I studied some period doubling bifurcations in systems of three degrees of freedom, but the problem became hopeless. I found hundreds of families of period-4 orbits. Thus, I decided to limit myself to families of period 1 or 2. Even so the problem is quite complicated (Contopoulos 1986a).

A particular complication is produced by certain orbits that we call "critical". Around such orbits the system is not "unique". That means that if we follow a family of periodic orbits along a path in the parameter space (ε, η), surrounding a critical point, we come back to the original point with a different orbit! In other words if two dynamical systems evolve slowly from one configuration (with $\varepsilon = \varepsilon_1, \eta = \eta_1$) to another (with $\varepsilon = \varepsilon_2, \eta = \eta_2$) following different paths in the parameter space they come to the same potential, but with a different structure of orbits (Contopoulos 1988a).

A similar nonuniqueness appears in the construction of self-consistent models, by using Schwarzschild's method. It was found that the same potential (i.e. the same density distribution) can be generated by two, or more, different distributions of orbits (Pfenniger 1984; Contopoulos 1988b).

The study of the orbits close to a complex unstable periodic orbit was a particular problem of interest (Contopoulos et al.1995a). In 2-D cases of stability and instability, orbits close to a periodic orbit intersect a surface of section along an ellipse, or a hyperbola respectively. But in the case of complex instability orbits near a periodic orbit intersect a surface of section along a spiral. These spirals can be given by a linear theory that agrees very well with the numerical calculations. However, as the orbits recede away from the periodic orbit, nonlinear effects limit their expansion and bring them back close to the periodic orbit, producing chaos.

The first time (1985) that I spoke about complex instability was at a meeting in Berkeley, California on "Orbital dynamics and applications to accelerators" (Contopoulos 1986c). The organizers paid my trip from Athens to California just for this meeting. I was supposed to fly by an American company. But in order to pay the cheapest fare, I flew to Germany and I took an American plane to San Francisco. Because of that I was penalized, i.e. they did not pay the European segment of my trip.

At any rate the meeting in Berkeley was very interesting. After my talk Dr. A. Lichtenberg came to see me and invited me to lunch. He told me that he was pleasantly surprised to hear my talk. "Usually, he said, the speakers at a meeting speak about a subject that they have covered, again and again, in previous meetings and publications. Your talk was the only novel and original lecture today". After that we had a long discussion with him and Dr. M. Lieberman at their department in the University.

Since that time I have close contacts with Dr. Lichtenberg. He invited me once more to Berkeley and we always exchanged preprints. I saw with great interest the book of Lichtenberg and Lieberman "Regular and Chaotic Dynamics" (1992), which is a very good introduction to this field.

An interesting approach to 3-D systems is by considering simple 4-D maps. These maps mimic the behaviour of a 3-D system on a Poincaré surface of section.

A particularly simple 4-D map is composed of two coupled standard maps. This map was introduced by Froeschlé (1972). The standard map (Eq. 41.2) is a well known 2-D problem that has been studied extensively up to now. But the coupling of two standard maps introduces new phenomena. Such a new phenomenon is the diffusion discussed by Arnold, that we mentioned above.

It is well known that Arnold diffusion is extremely slow. But if the perturbation is large, another type of diffusion takes place, namely resonance overlap diffusion (section 13).

I tried to find a simple method to distinguish between resonance overlap diffusion and Arnold diffusion. The coupled standard maps gave a very good example (Contopoulos and Voglis 1996). Orbits starting in a certain domain are confined in this domain for some time and then escape to a large "chaotic sea". The escape time (diffusion time) is larger when the coupling parameter β is smaller. If β is relatively large the diffusion is very fast. As β decreases the diffusion time increases exponentially in $(-\beta)$. This is well understood in the case of the resonance overlap diffusion. But when β becomes smaller than a critical value β_{crit} the diffusion time increases superexponentially, and this is an indication of Arnold diffusion. What is remarkable is the abrupt change of the time scale at the critical coupling β_{crit}. This seems to be the most vivid example of differentiation between resonance overlap and Arnold diffusion.

My work on 3-D systems continued in later years and a number of further applications appeared. However, this problem is far from exhausted.

29. BIFURCATIONS

The problem of successive bifurcations of families of periodic orbits attracted my attention since 1968 when I was visiting professor in Columbia and Harvard (section 15). Some years later I found a case of successive bifurcations in a simple dynamical system along the "central" family of periodic orbits. This family consists of straight line orbits, that become successively stable and unstable an infinite number of times as the energy increases up to the escape energy. The successive intervals between bifurcations have a ratio that tends to the number $\delta = 9.22$ (Contopoulos and Zikides 1980, Contopoulos 1981b). At every transition from stability to instability there is a bifurcation of a stable family of periodic orbits of equal, or double period, and at every transition from instability to stability there is a bifurcation of an unstable family. The stable bifurcating families have further period doubling bifurcations that lead to an infinity of bifurcated families.

At that time it was already known that the intervals between period doubling bifurcations in dissipative systems decrease by a factor $\delta = 4.67$ (Feigenbaum 1978, Coullet and Tresser 1978). This is a universal value, because it can be shown that it is the same in generic dynamical systems. Then Benettin et al. (1980) studied the period doubling bifurcations in a Hamiltonian system and found a ratio $\delta = 8.72$ (Fig. 25). They commented that this ratio, that was so different from the universal ratio $\delta = 4.67$, was close to the value found by Contopoulos and Zikides (1980).

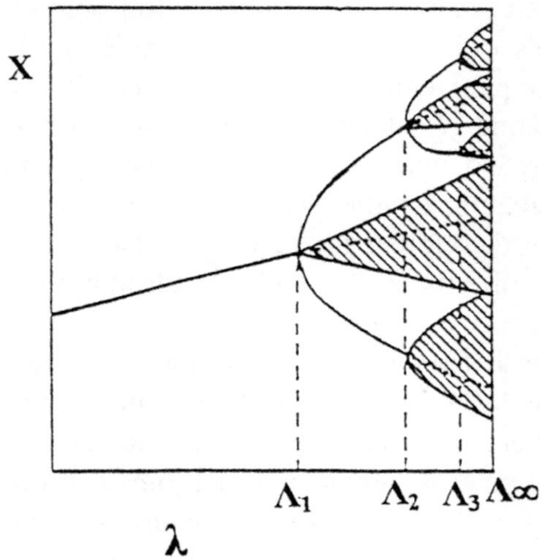

Figure 25. Period doubling bifurcations in conservative systems lead to a universal bifurcation ratio $\delta = (\Lambda_{i+1} - \Lambda_i)/(\Lambda_i - \Lambda_{i-1}) \to 8.72$.

However, it was later realized that the period doubling ratio $\delta = 8.72$ for conservative systems is also universal, i.e. it is the same for generic conservative systems, while the ratio $\delta = 9.22$ is not universal. It was proven by Heggie (1983) that the ratio $\delta = 9.22$ is equal to $exp(\pi/\sqrt{2})$. This ratio refers to bifurcations along the same family, and is different in other dynamical systems. We discussed these problems at a Workshop on bifurcations in ESO (Geneva) during 1980. Among the participants at that meeting were L. Galgani and J. Greene, who later referred to our discussions in their papers.

The complete story has some further aspects. A number of people had found independently the successive period doubling bifurcations. E.g. two Germans, Grossman and Thomae (1977) had found the ratio $\delta = 4.67$ in a dissipative system, but they had not established its universal character.

On the other hand I had calculated the successive bifurcations in a conservative system already in 1969 (Contopoulos 1970b). I had noticed that the intervals between successive bifurcations decrease with the order of the bifurcation. But I did not calculate the bifurcation ratio. Looking back at my notes I found that this number was, indeed, $\delta = 8.72$ (the universal ratio for conservative systems). Thus I missed an important discovery at that time.

I found recently a rediscovery of the ratio $\delta = 9.22$ (Brack 2001). I drew the attention of the author to the existing literature on this problem, and he thanked me for that.

Another aspect of the bifurcation problem was the discovery of inverse sequences of bifurcations (Contopoulos 1983). In some important cases the sequence of infinite period doubling bifurcations is followed (as the energy increases) by an inverse sequence, that leads to a single family, by joining all higher order families in pairs until only one family is left. This phenomenon was found first in a rotating galactic model, but afterwards it was found also in nonrotating systems, by me and, later, by several other people. The increase of the number of bifurcating families leads to an increase of the number of unstable periodic orbits, therefore to an increase of chaos. In the same way the decrease of the number of bifurcating families leads to a decrease of chaos, at least locally.

It is interesting to note that in some systems the increase of a control parameter (e.g. the energy) leads to large chaos, while a further increase of the same parameter leads to smaller chaos. Such cases seem rather exceptional. However, one can easily devise a control parameter ε that joins two limiting integrable systems, that exist for $\varepsilon = 0$ and $\varepsilon = \infty$ respectively. For small and for large ε the system is mostly ordered, while for intermediate values of ε the system may be mostly chaotic.

30. THE IAU GENERAL ASSEMBLY IN GREECE

After the end of my term as General Secretary of the IAU, in Grenoble (1976), I remained for three more years in the IAU Executive Committee as a consultant. My duties were limited to attending the meetings of the Executive Committee. The last meeting that I attended was at the Montreal General Assembly (1979).

When I arrived at the meeting room in Montreal I was a few minutes late. The former president, Leo Goldberg, welcomed me and said: "George, we just decided to have the next General Assembly in Greece". I looked in disbelief. "No kidding" I said. In fact I knew that the IAU had decided to hold the 1982 General Assembly in Bulgaria. But the president insisted. "We are not kidding. Look at the telegram in your file in front of you". It was a telegram from the Bulgarian National Committee stating that they would not be able to organize the 1982 General Assembly". Thus, the Executive Committee proposed Greece as an alternative.

I said that this was a very difficult matter that I had to discuss it with the Greek National Committee. I sent immediately a telegram, but I did not expect an answer soon. Two days later this problem was discussed again at a meeting of the national representatives. I explained again the difficulties of organizing a General Assembly in Greece. Then a number of other countries volunteered to take over this General Assembly. Nevertheless when a vote was taken about the best possible solution, a great majority voted for Greece.

During the Montreal General Assembly I had several discussions with people working in galactic dynamics. In particular I had many discussions with Martin Schwarzschild. On previous occasions I had emphasized to him the importance of calculating orbits in spiral and barred galaxies. This time I was impressed by the extent of the work he had done recently on this topic, quite analogous to my own work. Very soon Martin Schwarzschild became an expert in the field of orbits and integrals in galactic dynamics. And his authority played an important role in promoting this type of research.

After the Montreal General Assembly we discussed the possibility of organizing the next IAU General Assembly in Greece at the Greek National Committee. We decided to approach the Greek ministry to ask for support. Meanwhile a group of astronomers in Patras (Dr. Goudas and others) volunteered to take over the organization of the General Assembly. They approached some local politicians, who were enthusiastic with this idea, and promised to get the support of the ministry. Thus, we sent a letter to the new General Secretary of the IAU, Dr. Wayman, explaining the situation. But when the official answer of the ministry arrived it was very vague, and did not give us the required promise of support. In view of that we sent a negative reply to the IAU. With such a limited support we would not be able to organize the General Assembly. But then we received a telegram from the General Secretary: "Wait. I am coming to Greece". When Dr. Wayman came he contacted the astronomers in Patras and some of the local authorities, who promised to find support for the IAU. Finally, Dr. Wayman said that if the local money would not be sufficient, the IAU would subsidize the amount needed.

Under these conditions we reluctantly accepted to organize the 1982 General Assembly in Patras. There should be also two Symposia in Greece after the General Assembly, one in Crete on Cosmology and one in Thessaloniki on Dynamical Systems. I took over the organization of the Symposium in Crete.

The General Assembly took place at the new campus of the University of Patras outside the city, and, despite some organizational problems, it was quite successful. Besides the scientific meetings, there were several excursions, and social events, including a theatrical play "Iphigenia in Aulis" by Aeschylos.

One of the highlights of the General Assembly was the invited lecture by Y. Zeldovich on Cosmology. It was so lively and impressive that people did not stop clapping hands after the lecture until Zeldovich came back to the podium and said: "You are mistaken. I am not Iphigenia", and he added: "If you want to hear more about cosmology, come to the Symposium in Crete". I was astonished, because I had invited him to Crete and he had replied that he could not come. After his talk I told him: "So after all you are coming to Crete!". "I am sorry", he said: "I am allowed to go only to one meeting abroad. And that is good enough". Then he said in Russian "Harasho, harasho". That was one of the few Russian words that I understood. Then an Englishman who was nearby smiled. And Zeldovich turned to him. "Dr. Contopoulos and I understand very well each other because we are both Orthodox Christians".

This was the first time that Zeldovich was allowed to come to a western country. But he was not left alone. He was constantly escorted by a person who claimed to be an astronomer, but no one knew his name.

During the General Assembly Zeldovich was invited several times to dinner. At such a dinner someone asked him: "Who is this person that follows you constantly"? And Zeldovich replied: "He is my 007. But not so clever, as 007".

At the General Assembly there were a few more Soviet astronomers, among them Shklovsky, Kardashev and Einasto. They had been given just enough money for the General Assembly and the travel ticket. We had also provided full pension during the Symposium in Crete. But they had no money for hotels and meals for the time between the two meetings. Thus, I took them to Athens, where they stayed at the Observatory of Pentele, outside the city, and we provided also the meals. After that I sent them to Chania, Crete, to stay for three days, until the beginning of the Symposium, in the family house of my wife. An aunt of my wife lived there, and she covered all their needs. Our friends were very happy.

During their stay in Athens I had invited them, together with Zeldovich, to the University, and then to dinner. But Zeldovich came with his escort. I explained to him that I could take only four people in my car, thus his escort could not come with us. I suggested to take him to the nearest bus stop and give him instructions how to go to his hotel. But Zeldovich said: "You cannot separate us. We are like twins, like the stars in the constellation Gemini. If he does not come with us, I cannot come, either". Thus, finally I said: "I will place him as a third person in front, but when we approach a policeman he will have to lie down as much as possible, to be inconspicuous".

We had a very nice dinner on a hill outside Athens. Shklovsky was outspoken against the Soviet regime, as usual. But Zeldovich was clever to change the subject of the discussion, when it reached a dangerous point.

After that we had the cosmological meeting near Chania in Crete. It took place at the Orthodox Academy of Crete, a large building near the beach. We had a long lunch break that was used mainly for swimming and siesta. The participants were happy with this arrangement. We had also some excursions and social events during the evenings, during which the participants practiced Greek dancing.

The president of the organizing committee was Dr. Abell. During one excursion he left his clothes in the bus and he went swimming. But when he returned, this bus had left, and another bus had come. Dr. Abell was embarrassed, but he continued his excursion in his swimming trunks.

The last day of the symposium Skhlovsky was swimming most of the day with my children. When we asked him if he would come to the last session, he said with a sigh: "Oh, this is the last day! I want to enjoy it".

One year later Shklovsky and I were invited to a meeting in France. The meeting was very good, and the organizers did their best to make us happy. During the closing dinner everyone was praizing the organizers. But then Shklovsky got up and said candidly: "That is very good. But there is no comparison with the meeting of Contopoulos in Crete"!

31. BROUWER PRIZE

In 1982 I was awarded the Brouwer prize of the American Astronomical Society. This was a surprise for me. In fact until that time only astronomers in the USA received this prize. Later several distinguished Europeans, like Hénon, Fricke and Lynden-Bell received the same award.

Dirk Brouwer was a classical celestial mechanician. His main book (with G. M. Clemence) on "Celestial Mechanics" (1961) includes several recent developments in the field, like the von Zeipel method, and the dynamics of satellites, that give a quite modern aspect to the book. He was a pioneer in extending the scope of dynamical astronomy to cover also stellar and galactic dynamics. He invited a number of dynamical astronomers to Yale, belonging either to celestial mechanics or to stellar and galactic dynamics, and he encouraged them to expand the areas of dynamical astronomy. He organized also many meetings on modern aspects of dynamics. Such were the Yale summer schools on dynamical astronomy, that later became the Cortina summer schools, in northern Italy. He organized also several Workshops and Symposia, among them the IAU Symposium No 25 on "The theory of orbits in the solar system and in stellar systems", in Thessaloniki, Greece, that was the first large meeting that included stellar dynamics in it.

Thus, Brouwer can be considered as the father of modern dynamical astronomy and it is very appropriate that a prize in his memory was introduced by the American Astronomical Society.

The prize is usually followed by a lecture, either at a special meeting of the Dynamical Astronomy Division, or at the general meeting of the American Astronomical Society. I preferred the second alternative, and it was arranged that I should give a lecture at the Troy (New York) 1982 meeting of the AAS.

(Someone remarked that it was remarkable that a Greek received the AAS prize in Troy, of all places!).

My lecture was supposed to last about one hour. There were several hundreds of people attending my lecture and I wanted to give them a feeling about the importance of dynamical astronomy for astronomy and astrophysics in general.

In view of the solemnity of this occasion, I did something unusual for me. I wrote my speech, carefully selecting my words, and I started reading it, instead of speaking as usual with only a few notes in my hands. But I realized soon that my speech was not lively enough, and with the inevitable interruptions to show my viewgraphs, it would take a much longer time than scheduled. Thus, I put aside the written text and continued with only my slides to guide me.

That was a good decision, and it seems that everyone was satisfied. I think that my main message went through to the audience, that dynamical astronomy is not an esoteric subject of a separate group of isolated people, but an integral and very lively part of astronomy and astrophysics in general.

I described the relations between dynamical astronomy and other branches of physics, in particular high energy physics and elementary particles. I emphasized the mathematical aspects of the sciences of the Universe and concluded that the basic elements of the Universe are differential equations. After my talk Martin Schwarzschild, came to me and said laughingly: "George, when you mentioned the differential equations, I expected you to say that they describe the matter of the Universe. But you said that they are the matter of the Universe". And I replied in the same tone: "That is what it means to be a pure mathematician".

After the Troy meeting I made a grand tour all over the United States to give a series of lectures. I went to Boston (MIT), Washington (Naval Observatory), Chicago (University of Chicago), Urbana (University of Illinois), Tucson (Kitt Peak National Observatory), Texas (University of Texas at Austin), Atlanta (Georgia Institute of Technology), Tallahassee (Florida State University), Gainesville (University of Florida) and New York (Columbia University)

This tour was organized by some friends of mine, mainly by Dr. Peter Vandervoort of Chicago. Thus, I had an opportunity to visit many old friends and make new acquaintances.

In particular a quite new prospect opened up during my visit to Florida. The chairman of the Astronomy Department of the University of Florida, Dr. H. Eichhorn, volunteered to drive me from Tallahassee to Gainsville (3 hours drive). During this trip he asked me whether I would like to come as a visiting professor to the University of Florida. I was glad to accept. This offer was renewed every year. Thus, I started a very fruitful collaboration with the University of Florida that lasted for 11 years, from 1984 until 1994 (see section 38).

After the IAU General Assembly in Greece I returned to the United States for two months, as a visiting professor at Cornell University. I was invited by the chairman of the department, Dr. Yervant Terzian, and I gave a regular postgraduate course on Dynamical Astronomy. I had some excellent students in my class. I gave them, as homework, some small projects and their results were so good, that I included them in one of my papers, with the proper acknowledgements. I found the scientific environment in Cornell excellent from every point of view, and I was happy also that my son, John, did his thesis in Cornell.

32. VISITS TO ESO

After my first sabbatical at ESO (1976-1977), I was for many years a research associate there (1979, 80, 82, 83, 84, 85, 86, 87, 88). Besides these extended visits I visited ESO many times to attend meetings, or give seminars.

My association with ESO has been quite fruitful. I published there about 50 papers in journals or invited papers in several meetings. Most of them were first issued as ESO preprints.

ESO was very generous to me. It provided a good salary, a nice apartment, and extensive computing facilities. I was allowed also to invite colleagues from several countries to collaborate with me. Thus, I invited some of my assistants and associates from Greece, but also people from Italy, Switzerland, Germany, France and Japan. Furthermore, I organized a few Workshops and several meetings of the local ESO people and visitors. Two international Workshops took place in Geneva in 1980 one on "Orbits in Galaxies" and the other on "Bifurcations". I attended also many meetings all over the world, including a meeting in Santiago, Chile. ESO was in many cases providing my travel expenses.

The scientific environment in ESO was excellent. I had the opportunity to discuss all kinds of scientific problems with the staff and the visitors of ESO. Fifteen of my ESO papers were done in collaboration with other colleagues that worked in ESO for some time.

But for me a gift from heaven was the use of ESO's computers, with all their facilities, like the Midas system of graphics, the possibility to use colors, etc. I was doing several calculations at the same time, using every new computer that was available. Thus, finally, I generated a problem for the other users, who noticed that the speed of their computations was greatly reduced.

In ESO they used to have a party once a year. During such a party there was a play, written by an imaginative colleague, which made fun of various situations in ESO. One such situation was the following. Two persons work in the computer. One of them asks: "Why is the computer soooo slow today"? And the other replies: "Don't you know the Computopoulos effect"?

I was fortunate, because ESO was primarily interested in observational work, which did not need much computer time. Only a few people were doing theoretical work.

The Director General of ESO was Dr. L. Woltjer. He was very effective, especially in the establishment and operation of the big ESO telescopes in Chile[9].

Woltjer was wise in his administration of ESO. He gave ample freedom to his associates, but he expected definite results in return. He encouraged every positive prospect but he did not tolerate nonsense.

At the same time he initiated and supported many European activities and collaborations, like the European journal "Astronomy and Astrophysics", the European site of the Hubble space telescope, the ESO-Soviet schools for young astronomers, the ESO-CERN meetings on Cosmology and High Energy Physics, etc. He made ESO the center of the European Astronomy.

During my stay in ESO I was using a bicycle to go to my house. That gave me the opportunity to go around a lot, to the woods around Garching, and to picturesque villages further away.

Besides my several trips further away from Munich, I had the opportunity to go on excursions to many beautiful places in the Alps, palaces like Linderhof, Chiem-See and Neuschwanstein, the nice lakes of Southern Bavaria and several places in Austria, like Innsbruck, Saltzburg, Ramsau and Badgastein.

I have such nice memories from ESO and Garching that I am happy when I have the opportunity to go there.

A member of the staff, with a rather high position, is Dr. P. Grosbøl, a former student of mine. Grosbøl has been with me in Thessaloniki, Greece, and he did his thesis on the places of formation of the stars that are presently in the solar neighbourhood. During his stay in Thessaloniki he met a girl assistant in my department and later he married her. I did not know that (because I had meanwhile moved to Athens) until I met Grosbøl at a party in Garching. He told me that he was recently married, and I congratulated him. Then he added: "You know my wife, she is Barbara". I was happily surprised.

The recent work of P. Grosbøl is directed in finding indications of chaos in galaxies, a really very interesting endeavour.

[9]The history of ESO has been nicely exposed in the book of its first Director General, Dr. A. Blaauw under the title "ESO's Early History" (Blaauw 1991). ESO played a very important role as a rallying spirit in European astronomical activities. Furthermore the fact that ESO started in CERN facilitated the formation of ties with the physicists, that were realized, e.g., in organizing joint meetings with CERN, mainly on cosmological topics.

33. TERMINATION OF SPIRALS AND BARS

A subject that I studied with Grosbøl in ESO was the termination of the spiral arms of normal spiral galaxies.

I noticed that the periodic orbits in strong spiral galaxies are oriented in phase with the spirals, and enhance the spiral arms up to the 4/1 resonance, but they are completely out of phase beyond the 4/1 resonance (Fig. 26a). These orbits are stable and they trap around them many nonperiodic orbits. Thus, the spirals can be self-consistent up to the 4/1 resonance, but not beyond this resonance. My conclusion was that strong spirals should terminate at the 4/1 resonance. I published a paper in Comments on Astrophysics stressing this result (Contopoulos 1985).

Then Grosbøl and I decided to study this problem further (Contopoulos and Grosbøl 1986, 1988), using an extension of Schwarzschild's (1979) method to find self-consistent models of galaxies. Schwarzschild (1979) calculated orbits starting at a grid of initial conditions in a given potential, corresponding to an imposed density, and then he used a linear programming method to populate the grid points with appropriate numbers of stars in order to find the best agreement with the imposed density.

Instead of that we used the main periodic orbits at different distances from the center and an appropriate dispersion of velocities in order to find self-consistency. The various models depend on certain parameters, like the phases and the amplitudes of the spirals, the dispersion of the velocities, etc. These parameters are adjusted in order to find self-consistency.

In the best self-consistent models the spirals terminate at the 4/1 resonance.

More recent studies by Patsis et al. (1991, 1994) have considered also very different models, where the spirals terminate at corotation, or at the outer Lindblad resonance, but they are not even approximately self-consistent. Therefore, it seems that all strong spirals terminate at the 4/1 resonance. These are the Sb and Sc spirals. On the other hand weak spirals (of Sa type) can be extended all the way to corotation. Patsis, Hiotelis, Contopoulos and Grosbøl (1994) considered also the behaviour of the gas in self-consistent models. It was found that beyond the 4/1 resonance there are double weak extensions of gaseous spiral arms. This feature is, in fact, observed in many spiral galaxies (Fig. 26b).

This work has been extended now to many galaxies and the results converge to the same conclusion.

On the other hand the bars of barred galaxies seem to terminate at corotation, or a little before it. This was emphasized in a paper of 1980 (Contopoulos 1980). Many arguments show that bars cannot extend beyond corotation, and this view is generally accepted today.

Beyond corotation in barred galaxies there are spirals extending up to the outer Lindblad resonance. Self-consistent spirals have been found clearly beyond the outer 4/1 resonance, but the orbits of stars near corotation are chaotic.

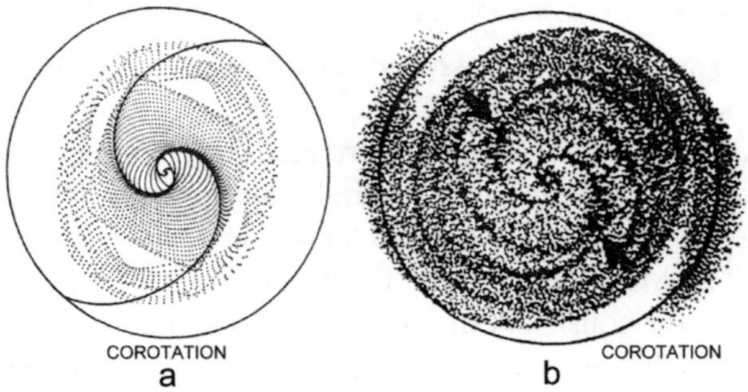

Figure 26. (a) Stellar orbits support the spiral up to the 4:1 resonance. (b) The gas also supports the spiral up to 4:1 resonance (arrows). Beyond it there are weak double gaseous arms up to corotation.

On the other hand the gas beyond corotation can join the ends of the bar with the outer spirals (Contopoulos et al. 1989). Furthermore even the chaotic orbits beyond but close to corotation tend to generate a larger density along the spiral arms, i.e. they contribute partly to self-consistency.

A detailed study of self-consistent bars was made by D. Kaufmann, who did his thesis on this subject under my supervision at the University of Florida (Contopoulos and Kaufmann 1992).

Another approach to the self-consistency problems of barred galaxies was advocated by Sellwood and Sparke (1988) and Sellwood and Lin (1989), who support the view that the spiral arms of barred galaxies outside corotation have a different pattern velocity. This is a plausible hypothesis and several people are presently studying this possibility.

34. FURTHER TRAVEL

In general my trips were rather complicated. They were not at all like the shuttle service between New York and Boston, that I used rather often, which resembles to a bus service. Instead, I usually had to change several planes and my schedule was rather tight. Sometimes the inevitable delays produced unpredictable changes in my schedule. On one occasion (1978) I was returning from Copenhagen to Athens. I should fly through London, but my plane was late and I missed my connection to Athens. It was late in the evening and I had an important faculty meeting next morning. I told my problem to the air company and they sent me to Geneva, late at night, put me in a hotel there, and reserved a seat in an early flight next morning to Athens. I arrived just in time.

On another occasion I had to fly from Hamburg to Leningrad (St. Petersburg) through Helsinki (1974). But there was a strike in Hamburg and the plane was late. When we arrived in Helsinki we saw the Aeroflot plane leaving (this plane was right on time!). What was I supposed to do? Fortunately it seems that I have friends in several places around the world. In Helsinki I knew professor Tuominen. I found his telephone and called him. He was kind enough to take me from the airport to the University dormitory. Next day I had several interesting discussions with his staff at the Observatory and then Dr. Tuominen arranged a very nice sightseeing tour for me in Helsinki and its surroundings before going to the airport. It was ironic that this time the Aeroflot plane was late by two hours. If it had the same delay the previous day I would not have to stay one night in Helsinki.

Dr. Tuominen felt an obligation to me because I had supported his candidacy for full professorship at the University of Helsinki. I was a member of an international committee, appointed by the University, to make recommendations for the chair of astronomy. There was one more serious candidate for this position, but I felt that Dr. Tuominen was better. I wrote a detailed report to support my view. There was another external referee who simply wrote in his report: "Both candidates are very good. So you may choose any one you like". Therefore, my opinion was essential for the senate of the University, which was to decide who should take the chair of astronomy.

After I submitted my report, I received a letter from a member of the senate, a professor of humanities, who wrote: "I see that you propose Dr. Tuominen. But I prefer the other candidate. Therefore, could you please send me a letter supporting him"? Of course I did not even reply to this letter, and Tuominen was elected.

Some of my trips were very long, like a trip to Australia in 1973, a trip to Chile in 1984, and a trip to Argentina in 1995.

The trip to Chile was particularly interesting. At that time I was in ESO that would pay my travel expenses. There was an office of Lufthansa in ESO that issued the tickets. They made arrangements that I should fly to Chile through Brazil. But when I saw the price I was surprised. "Is there no cheaper way to Chile"? I asked. "But you did not ask for a cheap fare", they said! Then they found that the cheapest itinerary was through New York. Thus, I flew to New York and I had just time enough to take the plane for Chile. But although I was just transiting, I realized that I had to pass through immigration in New York. I saw a long queue going through ropes that went around in a snakelike way several times. I realized that if I did not pass through immigration soon, I would certainly miss my connection. So I did something daring. I passed below several ropes and came to the front for the passport control. No one seemed to notice my rather irregular behaviour. When I boarded my plane I realized that my flight to New York was much shorter that the flight to Chile ahead of me. We went over Cuba, Venezuela, Equador, Peru and then a long way over Chile until Santiago.

We travelled all night long and next day we were still flying along the Andes, which were covered completely by snow. We reached Santiago in a wonderful winter day. I was to attend a symposium on "Nonlinear phenomena in physics" in a suburb of Santiago on the foothills of the Andes.

The Symposium was very interesting. There were lectures by M. Feigenbaum and P. Coullet who found the same universal scaling of successive bifurcations in dissipative systems independently of each other, by K. Thorne (on black holes, white holes, wormholes, etc) and by several physicists, working on statistical mechanics, cellular automata, quantum optics, chaos theory, convection, supergravity, Kaluza-Klein theory, quarks, supersymmetry and cosmology. I had many discussions with several people, both astronomers and physicists.

On one occasion I was talking in my hotel with a group of people about nonstandard statistical mechanics. Then one colleague introduced himself to me. I was surprised to hear him speak in Greek. He was Constantinos Tsallis, from Brazil. Tsallis is Greek, as his name indicates, but he lives permanently in Rio de Janeiro. His main work is on nonstandard statistical mechanics. He uses a more general concept of entropy (usually called Tsallis' entropy) that is applied to fractal sets. His ideas are extremely fruitful and many hundreds of papers refer to his work.

During the symposium I visited several times Santiago. It reminded me of the pre-war Athens. The same types of shops, the same tramways, the same open air shops, photographers and shoe-shining boys in the squares, the same old busses, etc. I had the feeling that I lived again in my past.

After the symposium the ESO people in Santiago arranged a visit to the ESO site in La Silla, north of Santiago. There was a small ESO plane, travelling from Santiago to La Silla. Then a bus was taking the visitors from La Silla to the Observatory at a height of 2500m. The plane was very small (about 8 seats) and it flew very low above the ground. I could see a huge mountain (part of the Andes mountains) in front of us, which we had to cross. When we approached the mountain the machines of the plane started roaring in a horrific way, and the plane flew higher and higher. I was afraid that any moment the plane would crash on the rocks in front of us. Fortunately the plane reached a pass, that it crossed, only a few meters above it, and then it went down again.

The airport of La Silla was a private one. No houses were around, except a small one-room wooden guard's station. But a bus was waiting for us. We went up the mountain to the Observatory that included at that time one 3.5 meters telescope and several smaller telescopes. Later, a second, new generation, 3.5 meters telescope was added.

The weather was very nice. Despite the snow that covered almost everything, the weather was hot, but crisp and the air was cool at the same time. I walked around for several hours. When I came back I found that my face and hands were extremely red, and burning. The sunburn was serious at such a high altitude.

Then we had lunch in the cafeteria. It was undoubtedly the best cafeteria in the world. All kinds of excellent food, fruits, ice creams, sweets, etc, all free of charge. Dr. Woltjer explained to me, later, that he had hired one of the best cooks in Chile, for the Observatory, to make the place attractive for the observers.

During the afternoon I visited the 3.5 m telescope and it was already operating in the infrared. They explained to me that the influence of the daylight on the infrared is very reduced, so that the telescope can operate most of the day. Later, at night, I visited various telescopes. I was impressed by the facilities available there. Dr. Dennefelt, from France explained to me the spectroscopic work he was doing. He did not even approach the telescope, but he was sitting in front of three computer windows. One was showing the field of stars seen by the telescope. The second was showing the spectrum (the lines of the spectrum became longer as more photons were gathered) and the third made a rough analysis of the data, strengths of the lines, etc, that were stored in the computer for further analysis. I remembered my early disappointment with our small telescopes in Greece and said that if I had such facilities, I could well become an observer myself.

Life was not always so happy at the Observatory. Sometimes the weather was so bad, that observations were impossible. But the time-schedule of the telescope was very tight. If one observer had bad luck with the weather, he had not the opportunity to do his observations a few days later. He had to re-schedule his programs for another period of time. I remember that my former student, P. Grosbøl, went for one week to Chile and he returned without making a single observation because the weather was very bad during this whole week.

Several years later I was invited to a meeting in La Plata, Argentina, in 1995, on "Chaos in gravitational N-body systems".

I flew there through London. There was a driver waiting for me at the Buenos Aires airport and he took me to the meeting place in La Plata. The distance seemed enormous. The roads were narrow but the driver went very fast. Finally after more than one hour we reached our destination. I met some Latin-American friends, which I knew already, like J. Muzzio, S. Ferraz-Mello and G. Sahade (former vice-president of the IAU) and made new friends there. I met also friends from the United States, and from other countries.

There were several new developments discussed there (Fig. 23). I spoke about the spectra of stretching numbers and helicity angles (section 45). The main new result that I presented was the clear distinction between Arnold diffusion and resonance overlap diffusion in 4-dimensional maps (section 28).

D. Merritt spoke about chaos in elliptical galaxies, especially around the central cusps. He had found that chaotic orbits form an important component in elliptical galaxies.

The younger colleagues of the La Plata Observatory were making a serious effort to adapt their research to the most recent developments in astronomy, especially in the field of dynamical astronomy. I made a comparison of this group with our group in Thessaloniki, Greece, several years ago, when we started working on this new field. Both groups were full of enthusiasm and had many ideas about future research. Therefore, I could predict with certainty the progress of my Latin-American friends.

On my way back from La Plata, I stopped in London and went to a Symposium in Cambridge, held in honor of D. Lynden-Bell on his 60th birthday. One of the main speakers there was de Zeeuw, who spoke about triaxial galaxies. He emphasized the work of Merritt on chaotic orbits in the central regions of elliptical galaxies. It was evident that the former attitude to consider only integrable models of galaxies was changing.

Another interesting trip was to the first Symposium in Florida (1984) on "Chaos in Astrophysics". The meeting was taking place in Palm Coast, on the Atlantic. They informed us that the best way to go there was by flying to Orlando and renting a car at the airport.

I arrived in Orlando late at night. It was morning of the next day by that time in Greece, and that meant that I was awake 24 hours already. According to the instructions I had to go to a particular rent-a-car company. I asked for a car but they wanted an account to charge, and I had no such account. Then they asked a huge amount of dollars as a deposit, much larger than what I was carrying. Fortunately some official understood my problem and accepted a rather small deposit. Then they gave me the key of a car and told me: "You will find the car in the parking place". "Wait a minute", I said. "I do not know how to locate the car". Then someone went and brought it to me. I gave him a tip and asked how the car operated. He gave me some instructions very fast, that I had no time to understand, and went away. Thus, I started experimenting with the car, to see the response of various buttons and levers. Finally I started. I was so tired, that I stopped to sleep a little. Finally I reached the Atlantic coast after 1 hour driving. I got out of the highway and I asked instructions about Palm Coast at a shop. But they had never heard of Palm Coast, although it was very close, and only suggested that I probably wanted Palm Beach, which was very far away. Finally I did find Palm Coast and reached my hotel completely exhausted, well after midnight. The hotel was on the Atlantic coast and had a nice view of the ocean, with its huge waves.

I met several friends there, including E. Spiegel from Columbia and L. Galgani from Milan. This meeting was the first on chaos in astronomy and astrophysics. It was the time when people started to realize that chaos was important, and many people were working on it.

Whenever I visited the United States I travelled a lot by plane all over the country. Sometimes I had to use very small planes, that could be quite bumpy. Such were some of my flights to Ithaca, New York in order to reach Cornell University. When I complained about such a flight, a friend told me that the earlier flights were even more primitive. They were not guided by radio, but the pilot had in front of him a map of the area and was following the main highways to reach Ithaca. The Ithaca airport was a meadow and deer were grazing there. The pilot had to go close to the ground to scare the deer and only after a turn he could land his plane.

35. ORDER AND CHAOS

The subject of order and chaos in dynamical systems absorbed my main interest over the last 50 years.

In the past it was generally believed that dynamical systems are either integrable, or ergodic. This general belief was explicitly stated in the first edition of the classical book of Landau and Lifshitz "Mechanics" (1976).

Integrability represented order, while ergodicity represented chaos. Celestial mechanics was a paradigm of order, while a paradigm of chaos was statistical mechanics. The motions of the planets and satellites were considered as typical examples of order, while the motions of atoms, or molecules, in a gas were considered as typical examples of chaos. For historical reasons stellar dynamics followed the methods of statistical mechanics. Thus, stellar motions were considered a priori to be chaotic.

However, gradually this simple classification changed. A precursor of this change was Poincaré, who in his "Méthodes Nouvelles" had explored, a little before 1900, the possible appearance of chaos in celestial mechanics (nonintegrability of the restricted three-body problem, and the topology of the homoclinic and heteroclinic intersections).

Only some rather special systems, like the Stäckel potentials and the Toda lattice, were found to be really integrable. Similarly, only some special maps (like the baker map) and some systems with abrupt reflections of the orbits (like the stadium) were proven to be really ergodic.

The situation changed dramatically in the late 50's and in the 60's. The main reason was the use of computers. The first computer calculations have shown that order and chaos coexist in most dynamical systems.

On one hand some systems that were considered chaotic were found to possess much order. The first example of this type was the Fermi-Pasta-Ulam paradox (Fermi, Pasta and Ulam 1955). These authors considered a system of N particles along a string, attracted by linear and non-linear forces, and expected to find equipartition of the energy between the various modes. In fact, this system can be reduced to N coupled oscillators, where the coupling is provided by the nonlinear terms. Because of the coupling it was expected that energy would be exchanged chaotically between the various modes. Instead, it was found numerically that the system was quasi-periodic and only a few modes were excited in an almost periodic way. This was a paradox, which excited an extensive activity in this field. It was later found that if the energy exceeds a certain critical value, chaos does appear in the system, and much work has been done in finding this critical value of the energy.

Another early example of order, in a presumably chaotic stellar system was found in 1956, by calculating orbits in a 3-D galactic model (Contopoulos 1958). This was explained later by a third integral of motion (Contopoulos 1960) (section 5). In this case again there is a critical value of the energy beyond which chaos is predominant. The critical value was first found by Hénon and Heiles (1964). For small energies most of the orbits are ordered, but for large energies most orbits are chaotic. The transition from order to chaos is quite abrupt. Namely at a value of the energy $h = h_{crit}$, chaos begins to grow considerably, and at a little larger h chaos becomes predominant. However, chaos is not complete. Even for h larger than the escape energy small islands of stability persist and they decrease only slowly (Contopoulos and Polymilis 1987).

This phenomenon is quite general. I.e. generic dynamical systems are mostly ordered for small energies, and as the energy increases beyond a critical value, they become mostly chaotic within a small interval of energies. But small islands of stability persist for larger energies.

The existence of order for small energies is the content of the famous KAM theorem (after Kolmogorov, Arnold and Moser), while the transition from order to chaos is implicit in the Nekhoroshev theorem. The question is now whether chaos becomes complete for even larger energies. There are now theoretical considerations (Newhouse 1983; Duarte 1994) and numerical evidence (Contopoulos et al. 1994a) that small islands of stability appear for arbitrarily large energies. Namely there is no interval Δh without any islands of stability. However, these islands are so small, that the system is practically completely chaotic for sufficiently large energies.

The most chaotic systems are called Anosov systems. Such systems do not have any islands of stability. They have an infinity of periodic orbits, but all of them are unstable. They have everywhere a finite Lyapunov characteristic number, i.e. exponential deviations of nearby trajectories (section 28).

A friend of mine from Atlanta, Joe Ford (Georgia Institute of Technology) was always emphasizing the importance of chaos in generic dynamical systems. He invited me several times to Atlanta for lectures and discussions, and to a number of meetings, and in one case I invited him to Greece. In another case he invited me to Berkeley, California, where he was a visitor (1979). I flew there from Austin, Texas, and I met also a number of specialists on chaos, like B. Chirikov (from USSR) and local specialists like A. Lichtenberg and A. Kaufmann. It was an informal, but very lively meeting.

But later I had a serious disagreement with Joe Ford. He published a paper on "algorithmic complexity" (Eckhardt, Ford and Vivaldi 1984), in which he claimed that he could find the properties of chaos by calculating orbits in rational polygons. The orbits are piecewise rectilinear and complexity is due only to the abrupt reflection of the orbits. Such systems are exceptional and they lack the most important properties of generic nonlinear systems, like Smale horseshoes (Smale 1963, 1967). J. Ford asked my opinion and I wrote him my reservations.

Unfortunately Ford not only disagreed with me, but he felt offended. He wrote me that others had appreciated his work and he considered it as the best way to understand chaos. I am sorry but I cannot agree. It is like trying to understand the properties of real numbers by only studying the rationals.

In stellar dynamics it was believed that most systems are ergodic, except for some exceptional cases. The situation was well described by Lynden-Bell (1998): "To set the scene, I had written my thesis in 1960 which contained a new derivation of what potentials had local first integrals of the motion besides the energy and the angular momentum about the axis... . Having classified the special forms of potential that had local integrals, I expected that most other potentials would show ergodic behavior. From the inequality of the z and R dispersions of the stars in the Galaxy it was clear that there must be another integral other than E and h for the Milky Way, so I began trying to fit Eddington [16] (now called Stäckel) potentials to galaxies. It was quite shattering when at the 1961 IAU general assembly in Berkeley, George Contopoulos [10] showed that orbits in most smooth potentials behaved as though there were third integrals".

In fact, it was realized that many real systems, like galaxies, are close to integrable. The usefulness of (nearly) integrable systems in galactic dynamics was emphasized by Schwarzschild (1979), who constructed self-consistent models of galaxies, which can represent real galaxies, without any chaos. Furthermore several people could fit integrable (Stäckel) potentials to real galaxies (Hori 1962; de Zeeuw 1985; de Zeeuw, Hunter and Schwarzschild 1987, and others).

In particular Schwarzschild was very much interested in formal integrals like the "third integral". He contributed significantly in changing the attitude of astronomers as regards the third integral. While in the 60's and 70's many people were still sceptical about such integrals, in the 80's most people believed in the ubiquitous existence of integrals of motion of the third integral type. The success of Schwarzschild's method made several people believe that chaos is unimportant, or nonexistent, in real galaxies. It was even stated that, somehow, the formation process of galaxies takes care to produce only ordered, or mostly ordered systems.

But chaos does exist in at least two regions of galaxies, the center and the corotation region.

Orbits in a tri-axial flat galaxy that go near the center are mostly chaotic. In fact, stars moving in such orbits try to follow a nearly Keplerian (point mass) potential near the center, while they try to follow some Stäckel potential far from the center. Both limits (Keplerian and Stäckel) are integrable, but their combination is not. Thus, the interaction of the two types of potentials produces chaos. Such chaos is predominant in galaxies with a black hole, or a density cusp, at the center.

The corotation region also produces much chaos in galaxies. In fact near corotation there is an interaction of many resonances that introduces a large degree of chaos there (Fig. 27).

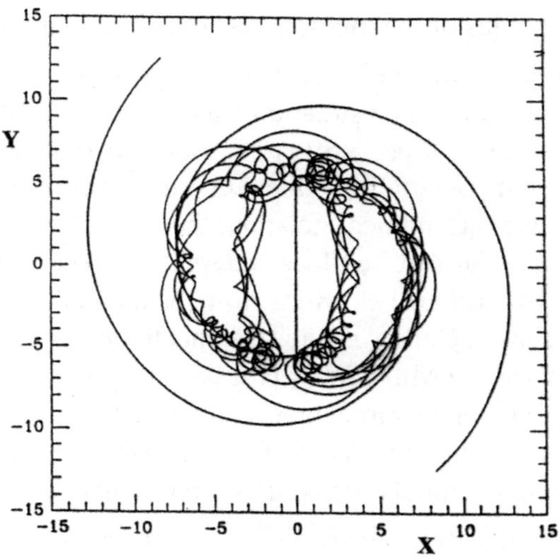

Figure 27. A chaotic orbit near corotation in a barred galaxy.

I always tried to keep the balance between studies of order and chaos in galactic dynamics. In the years after 1960, when most people believed in ergodicity in galaxies, I emphasized the near-integrability and orderly nature of most galactic systems. Later on, when the notion of order was well established, I emphasized the regions where chaos was important.

During a Workshop at the University of Florida in 1985, Vandervoort made the following remark. "George Contopoulos has emphasized for many years the integrable aspect of galaxies. Now he wants us to turn around and see the chaotic aspect of these systems. It is like getting double mileage for the same money".

I am glad to say that this double aspect of galactic systems is now generally accepted (e.g. Merritt 1999). Our recent work (Contopoulos, Efthymiopoulos and Voglis 2000; Voglis et al. 2002) on self-consistent models of galaxies, generated by N-body simulations, emphasizes both order and chaos in galaxies. Order and chaos are necessary in order to construct realistic self-consistent models of galaxies.

Recently a book of mine on "Order and Chaos in Dynamical Astronomy" (Contopoulos 2002) was published by Springer Verlag (Fig. 28b) (see section 55).

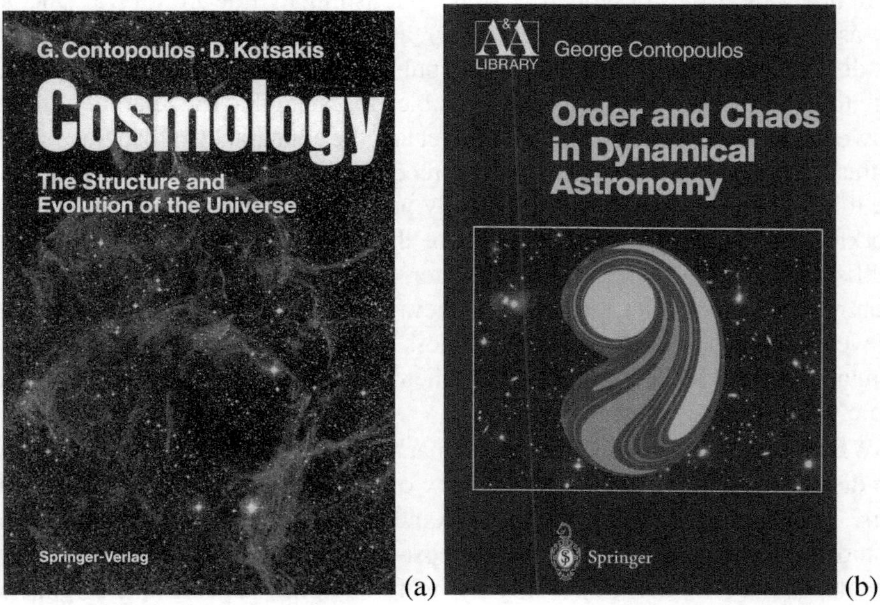

Figure 28. Covers of my books on "Cosmology"(*a*) and on "Order and Chaos in Dynamical Astronomy"(*b*).

36. RELATIVITY AND COSMOLOGY

During my stay in Chicago in 1963 I had started a third paper related to the post-Newtonian approximations of General Relativity, namely an application to the restricted three-body problem. I completed this paper, after my joint papers with Chandrasekhar, and it was submitted by Chandrasekhar to the Royal Society in 1967. However, meanwhile, a student of Chandrasekhar, Mr. Krefetz, had worked on this problem, and had published a paper in the Astronomical Journal, in May 1967. Thus, there was an overlapping of my paper with Krefetz' paper, and Chandrasekhar had to withdraw my paper. As he informed me afterwards, Mr. Krefetz had come to Yerkes after my departure and Chandra had suggested to him the restricted problem as an application of the post-Newtonian approximations. Krefetz finished soon this work and published his results after my paper was submitted (11 March 1967) but before it was considered for publication. Chandrasekhar wrote me (1 Nov.1967) "I am sorry that I did not realize how nearly Krefetz had duplicated your work". And he finished his letter with the words "please forgive me".

I learned a lesson at that time. You should publish your results as soon as possible, because there may be others that work in the same field, and they might precede you.

In many cases the consequences of not publishing in time are severe. One of my assistants, Dr. Bozis (later professor), in three cases missed the priority of his discoveries because of his neglect to publish fast. I myself took, in general, care to publish my results as fast as possible. I noticed that the interval of time between a lecture of mine on a new subject and the appearance of papers related to this lecture decreased considerably in recent years. After my first lectures on the third integral many people, especially younger research associates, started working in this field. But it took more than two years before they started publishing on this subject. However, later, this interval was reduced to a few months. Therefore, after I developed a new idea and gave some lectures about it I went to publish my results. Nevertheless I never denied my help and support to many people who later thanked me in their acknowledgements (and to several more who did not even thank me).

When I was working with Chandrasekhar on post-Newtonian approximations we discussed several times the possibility of extending our work to higher orders. The algebraic work involved was tremendous. In fact Chandrasekhar with a student worked out later the second post-Newtonian approximation (Chandrasekhar and Nutku 1969). Only by using computer algebra one might be able to go to higher orders. I was very enthusiastic with such methods, because I had just developed computer algebra routines for galactic dynamics. But then we had the problem of gravitational radiation that appears in the 2 1/2 post-Newtonian approximation. Because of this radiation the usual constants of mechanics, e.g. the energy, momentum and angular momentum, are no more preserved.

Unfortunately, when I had understood the problem, and we should begin serious work on it, I had to leave Yerkes. When I came back, a few years later, Chandrasekhar had already worked out the 2nd and 2 1/2 post-Newtonian approximations and he had derived the radiation reaction, a most important result, that refers to the loss of energy (and also momentum and angular momentum) due to the gravitational radiation.

But later we found another topic of common interest, namely "orbits in the problem of two fixed black holes". S. Chandrasekhar had studied the scattering of radiation from two fixed black holes (charged black holes, in which the electrostatic repulsion is counteracting exactly the gravitational attraction; Chandrasekhar 1989). He had calculated the absorption and scattering of gravitational and electromagnetic radiation by the two black holes. In an appendix he had included a few null geodesics, representing orbits of photons in the field of the two black holes.

He asked me if I would like to explore the subject further. I liked this subject and I made several theoretical and numerical investigations of orbits of photons (null geodesics) and of particles of non-zero rest mass (time-like geodesics).

The most interesting result was that, while the corresponding classical problem is integrable, the relativistic problem is chaotic. In particular a beam of photons coming from infinity is split into three sets (Fig. 29). Photons falling on the first black hole (I), photons falling on the second black hole (II), and photons that escape again to infinity (III). Between two orbits of two different sets there is always an orbit of the third set. Thus, between an orbit falling on the first black hole and an orbit falling on the second black hole there is an orbit escaping to infinity, etc. This is a clear demonstration that the three sets of orbits have a fractal structure.

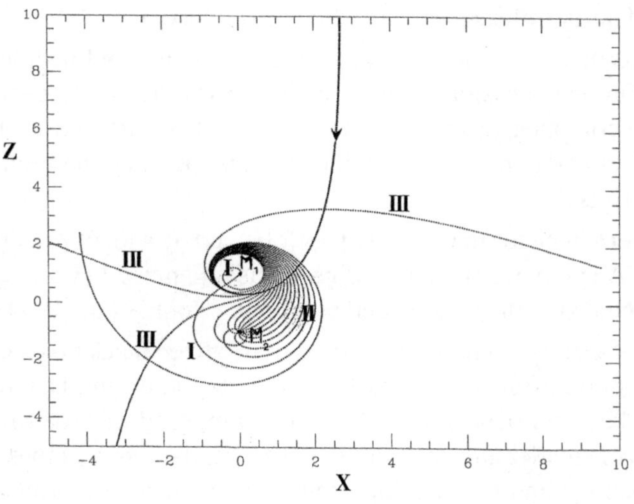

Figure 29. A thin beam of particles from above is split into three fractal sets of particles, reaching the two black holes I and II. or going to infinity III.

On the other hand in the case of particles that do not escape to infinity there is both order and chaos. We can prove the existence of ordered orbits that are trapped around stable periodic orbits, and chaotic orbits that fall into one of the two black holes, or are restricted in certain chaotic domains for all times.

This problem seems to have been the first where chaos was clearly demonstrated in general relativity. In particular it was found that the transition from order to chaos follows the same rules, as in other problems of classical mechanics. Namely a stable periodic orbit becomes unstable at a period doubling bifurcation, the bifurcated orbits become unstable at a period-4 bifurcation, and so on. The intervals between successive bifurcations decrease approximately geometrically with a ratio $\delta = 8.72$, which is the same universal constant, as in many other dynamical problems (section 29).

Chandrasekhar presented my first two papers on this subject to the Royal Society (Contopoulos 1990b, 1991). After that we had many interesting discussions on chaos.

One byproduct of my work on two fixed black holes was that there are no satellite periodic orbits that close around only one black hole in the classical case. This classical result had escaped notice, although many people had explored the classical problem of two fixed centers.

I was surprised that I could not find numerically such satellite periodic orbits. Thus, I looked more carefully at this problem and found an analytic proof that no such orbits exist. On the other hand in relativity the gravitational force is stronger and families of satellite periodic orbits do exist.

I mentioned this classical result in several colloquia, and people could not believe me. On one occasion a colleague in Texas told me: "I cannot believe you. I have a computer program for the two fixed centers and I will check it myself". The next day he came convinced, after spending the night in efforts to find such orbits.

The discovery of chaos in relativity was followed by a lot of research activity on this topic. A symposium on this subject was organized in Calgary, Canada, in 1994 and the Proceedings appeared as a book (Hobill et al. 1994).

A more recent study of chaos in a system of two fixed black holes refers to the basins of attraction of the two black holes (Contopoulos and Harsoula 2004). These basins have a fractal form. They are composed of broad regions and thin filaments of smaller and smaller thickness. At the limits of these filaments we find homoclinic and heteroclinic orbits (section 43), i.e. orbits that reach asymptotically the periodic orbits after infinite rotations close to them.

What is remarkable is that the black holes act as attractors, in the same way as the attractors of dissipative systems, although the system is conservative. The reason is that most orbits reach the black holes as singularities. As a consequence the orbits cannot be continued beyond these singular points. Thus most orbits are absorbed by the black holes and there is no conservation of areas as in general conservative systems. In this respect the situation is similar to the case of escapes (section 40), where again we have nonconservation of the areas on a surface of section.

The dynamical systems with singularities is a new topic that has probably many interesting prospects for research.

Another related topic was chaos in cosmology. This subject had a complicated development. A particular cosmological model, where chaos seems to be prominent, was the Mixmaster model introduced by Misner (1969), and independently by Belinskii and Khalatnikov (1969).

The name Mixmaster was introduced by Misner to indicate that there was a mixing of the orbits in the very early universe. If one replaces the cosmological time t by a new time, $\tau = -\ln t$, the big bang time $t = 0$ corresponds to an infinite τ-time. The Mixmaster model has three different expansion rates for the three space coordinates x, y and z. Two of them are in general positive (or negative) and the third negative (positive). Thus, e.g.,the universe expands along the axes x and y and contracts along z. But the expansion, or contraction, changes in a chaotic way along different directions. Therefore, later, the universe expands along the x and z axes, and contracts along the y-axis, and so on.

The first indications were that the Mixmaster model is chaotic. Some rough calculations indicated that the Lyapunov characteristic number (Eq. 28.1) is positive, and this is the basic characteristic of chaos. However, more accurate calculations later have shown that the Lyapunov characteristic number of orbits in the Mixmaster model is zero. This has been verified many times. Thus, people started wondering whether, after all, the Mixmaster model might be integrable.

At this point I, and two colleagues from France, Grammaticos and Ramani, attacked this problem from a different point of view. We checked whether the Mixmaster model satisfied a well known criterion for integrability, the Painlevé criterion. Namely if all the solutions are expanded in powers of the time t and have only poles as singularities in the complex plane, then the system is "probably" integrable. (In fact no counterexample has been found to this conjecture). What we found (Contopoulos, Grammaticos and Ramani 1993) was a most general solution, depending on six arbitrary constants, that satisfies the Painlevé condition. That was an indication of integrability. But we could not find any new integrals, besides the energy.

Later it was realized that there are other solutions of the Mixmaster model that do not have the Painlevé property (Latifi, Musette and Conte 1994; Contopoulos, Grammaticos and Ramani 1995b). This is a strong indication that the Mixmaster model is chaotic[10]. But the discussion continued. Two recent papers appeared under the titles: (1) "The Mixmaster model is not ergodic" (Cushman and

[10]In fact the Painlevé criterion for integrability requires that all solutions should have the Painlevé property.

Sniatycki 1995 and a more recent preprint), and (2) "The Mixmaster model is ergodic"(Cornish and Levin 1977).

Both these papers are correct from their respective points of view. The first paper emphasizes the fact that almost all orbits in the Mixmaster model escape to infinity. Therefore they cannot be ergodic. The second paper emphasizes the fractal nature of the orbits. At a meeting in Moscow in 2000 a participant presented a paper under the title "Contopoulos' paradox". The paradox was my opinion that the opposing views above are both correct.

The problem is clarified if we notice that the Mixmaster model is a case of chaotic scattering, but not chaotic in the usual sense, which implies a finite phase space. This fact explains why the Lyapunov characteristic number is zero, despite the fact that most orbits are chaotic. In fact, in a chaotic scattering case, two nearby particles escaping to infinity deviate from each other only linearly in time ($\xi \approx ct$ as $t \to \infty$) therefore their Lyapunov characteristic number (LCN) is zero (Eq. 28.5).

But when the particles are in a finite domain the short time LCN (section 45) is positive and the orbits are chaotic. A more recent discussion of the Mixmaster problem was made by Contopoulos, Voglis and Efthymiopoulos (1999). We studied the source of chaos in this model. This is to be found in the encounters of the orbits with the central part of the system, where large deviations of nearby orbits are generated.

Cosmology is a subject of particular interest that attracts the imagination of the students. As a student I had read a lot of books and articles on cosmology. As a professor I have given a course on general relativity and cosmology since 1957, which was always well attended by the students. Later, in 1982, I decided to write a book on cosmology for the benefit of my students, with my colleague, professor D. Kotsakis. The third edition of this book appeared in 1986. This book is at an intermediate level between a popularizing exposition and a mathematical textbook. It has been used extensively as an introductory course in cosmology at the advanced undergraduate level.

Then Springer Verlag decided to publish a translation of this book in English under the title "Cosmology. The Structure and Evolution of the Universe" (Contopoulos and Kotsakis 1986)(Fig. 28a).

The first part of this book contains the available observational results. The second part includes chapters on general relativity and on the early universe. It presents the ideas and concepts of modern cosmology with little mathematics, including the impact of high-energy physics on cosmology. The third part deals with some philosophical aspects of modern cosmology, like the arrow of time, the universality of physical laws, inflation and causality, and the anthropic principle.

The translation of the book was made by Dr. M. Petrou, a former assistant of mine, and her husband, Dr. P.L. Palmer from the University of London. The translation was perfect, and I had only to correct a few special expressions.

Several reviews about this book were published in various journals. E.g. the well known English relativist M. Rowan-Robinson wrote a review in Nature (1988) under the title "Greek Independence". His text explains this slightly strange title. It says: "this is a difficult moment to be bringing out a new book on cosmology. Contopoulos and Kotsakis's cosmology is a translation of a revised and updated version of their 1982 book in Greek. It survives the pitfalls of this fashion conscious era by pursuing a rigorously independent-minded attitude to contemporary ideas". "The third part is an excellent and profound discussion of the fundamental problems of cosmology: the universality of the laws of nature, the uncertainty principle, causality, the origin of time and time's arrow, teleology, the anthropic principle, metaphysics. For anyone with a philosophical turn of mind this section alone makes the book essential reading".

Similar comments were expressed in Soviet Astronomy (1988) by Dr. V.G. Surdin. "It is precisely in third part of the book, it seems to me, that the fact that its authors are Greek scientists is manifested more clearly. Not all books at this level are distinguished by such a breadth of view on the subject and a fundamental level of statement of the problems. Those readers who are lucky enough to become acquainted with modern cosmology from the book by Contopoulos and Kotsakis become fully aware of the integral character of this science, sense the enormous range of its language, from tensors and differential equations to biological and philosophical concepts, and catch the true grandeur of this science of the universe. A translation of the book into Russian would be successful in our country".

Finally I mention the comments of a German author, Dr. E. Schmutzer in General Relativity and Gravitation (1989). "When I first saw this book, I immediately realized that just this kind of presentation of the up-to-date results of modern astrophysics and cosmology is excellently suited for the use of relativists researching in this field". "I read this inspiring book with great pleasure and recommend it to relativists as a very informative source for their current research".

This book has been the first astronomical book that was translated from Greek to English.

Today this book is partly outdated by the recent dramatic developments in observational cosmology. However, I was happy that even at the end of 2003 I received a letter from a specialist (Dr. A. Kosowsky from Rutgers University) stating among other things: "You also touched on many topics that today have developed into the important questions in the field, so I think that you had a lot of foresight when writing the book".

37. QUANTUM MECHANICS VS CLASSICAL MECHANICS

Many people consider quantum mechanics as non-deterministic. That is wrong. The basic equation of quantum mechanics is Schrödinger's equation, which is equally deterministic, as are Hamilton's equations of classical mechanics. The only difference is that Schrödinger's equation is a partial differential equation dealing with waves, while Hamilton's equations are ordinary differential equations dealing with particles. It is well known that one can derive Schrödinger's equation by replacing certain functions in the Hamiltonian by appropriate operators.

The indeterminacy of quantum mechanics arises when we need to describe particles instead of waves. It is due to the dual nature of particles that are at the same time waves.

My interest in quantum mechanics dates from my student years, although there were no courses on this subject at that time. In fact I was the first to give a course on quantum mechanics at the University of Thessaloniki (1970).

I was particularly interested in comparing classical and quantum mechanics. A particular aspect of this problem is the semi-classical theory that deals with highly excited atoms, like Rydberg atoms, where classical results are good approximations of quantum results. There is a paper on this topic by User et al. (1991) under the characteristic title "Celestial Mechanics on a Microscopic Level".

Starting in 1977 I had some discussions on these problems with I. Percival and his group at Queen Mary's College in London. I had also many contacts with T. Uzer, at the Georgia Institute of Technology in Atlanta, and with Martin Gutzwiller, in New York. I consider M. Gutzwiller's book "Chaos in Classical and Quantum Mechanics" (Gutzwiller 1990) as an excellent introduction to this area.

But my deeper involvement in these problems started as a collaboration with Dr. S. Farantos, a professor of chemistry at the University of Crete, that resulted in a number of joint papers.

A classical particle is represented in quantum mechanics by a group of waves. The average position and momentum of such a group of waves, <q> and <p>, follows for some time a classical trajectory. Later, however, the waves are dispersed, and the wavefunction (which is given initially as a Gaussian wave packet) spreads over a large part of phase space. From this wavefunction one can derive, numerically, the spectrum and the eigenfunctions of the system.

In a formulation of quantum mechanics, due to Feynman, the wave at any point can be found by superimposing all possible classical orbits from the original point to this point. Gutzwiller (1990) noted then that the contributions of most orbits cancel by phase mixing, and the only orbits that contribute to the wavefunction are the periodic orbits. Some people (e.g. Davis and Heller 1981) then noted that one or a few periodic orbits play the most important role in this connection.

Our group (Founariotakis, Farantos, Contopoulos and Polymilis 1989) studied this problem as follows. We considered a well known classical problem of two coupled oscillators and calculated the most simple periodic orbits in it. Most of these orbits correspond to particular resonances. The same system considered quantum mechanically, gives a well defined spectrum, and each resonance is represented by a particular eigenfunction. We then compared the contours of the eigenfunctions with the corresponding periodic orbits, and found that the maxima and minima of the eigenfunctions are almost exactly on the corresponding periodic orbits. That is true also in the case of more complicated eigenfunctions, whose similarity with periodic orbits is not a priori evident. Only by comparing these eigenfunctions with the appropriate periodic orbits we can see the underlying similarity (Fig. 30).

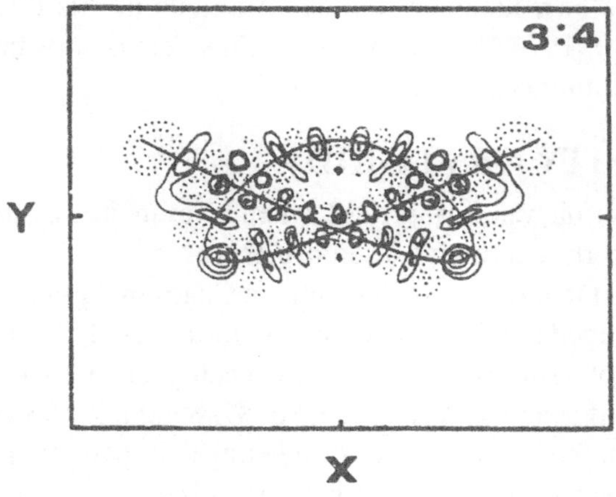

Figure 30. A quantum mechanical eigenfunction having its maxima and minima on a periodic orbit.

It is remarkable that this correspondence exists also for unstable periodic orbits. These are the so-called "scars" in the quantum mechanical terminology (Heller 1993).

Of course we do not explain all the details of the eigenfunctions, which are due to higher order resonances. But the basic similarity of the eigenfunctions with the periodic orbits is impressive.

We considered applications of this method to more complicated potentials, like the Morse potential (Farantos, Founariotakis and Polymilis 1989), the effects of various types of instability (Contopoulos et al. 1994c), etc.

More recently we considered applications to realistic models of molecular physics, like the water molecule (Contopoulos and Efstathiou 2001).

A problem that is still open is the problem of quantum chaos. It seems that quantum mechanical systems are much less chaotic than their classical counterparts. This is strange, because people would expect more chaos in quantum mechanical systems, due to the Heisenberg indeterminacy.

The explanation is that the waves of quantum mechanics are not influenced by the details of the classical motions, where we have chaos mainly close to unstable periodic orbits. Therefore if a KAM torus develops holes and becomes a cantorus (section 46), the corresponding waves "do not see" these holes and the corresponding transition to chaos.

This point of view is generally accepted today. However some people (Ford et al.1991) claimed that there is a "breakdown of the correspondence principle of quantum mechanics". But this extreme point of view is not accepted by most people.

Another important problem refers to a possible generalization of quantum mechanics to include also nonlinear terms. Attempts in this direction have been made by Penrose (1976,1992) and others. This field is at its first stages of development but its future is great (section 56).

38. UNIVERSITY OF FLORIDA

My first visit to Florida was during 1982, when I was invited by the chairman of the astronomy department to give a seminar there.

I came back in 1984 to attend a Symposium on Chaos in Palm Coast (section 34). Then it was arranged that I should come to Florida regularly, as a "visiting graduate research professor" for some months each year. I have been at the University of Florida during the years 1985, 86, 87, 89, 90, 91, 93 and 94. Furthermore, I have been there at a number of meetings. In particular my friends and colleagues of the astronomy department organized a special meeting on "Nonlinear dynamics and chaos in astrophysics" during 1998, and the proceedings were published as "A Festschrift in honor of G. Contopoulos" (Buchler et al. (eds), Annals New York Acad. Sciences 867, 1998).

During every lengthy visit I was giving a regular graduate course on nonlinear problems, plus a series of lectures on new developments in dynamical astronomy (once a week) that were attended mainly by faculty members. I was working on several topics, alone, or with other members of the staff, or graduate students, like Heinz Eichhorn, Steve Gottesman, Jim Hunter, Henry Kandrup, Howard Smith, Martin England, Dave Kaufmann, Elaine Mahon and Christos Siopis.

During my stay in Florida I published almost 50 papers in journals, or invited papers in conference proceedings. More than 10 papers were written in collaboration with members of the astronomy department of the University of Florida. Furthermore, two graduate students did their theses under my supervision. The first thesis was by Elaine Mahon. The chairman of the department wrote me about that: "I know you have taken a great deal of time and put in much effort on Elaine's work. ... I believe that it is one of the better pieces of work to come out of the PhD program in many years".

The second thesis was by David Kaufmann, who wrote in his introduction: "Dr. G. Contopoulos, the de facto advisor of my thesis, not only has taught me essentially all that I know about the fascinating topics of nonlinear dynamics and chaos, he has also provided an unmatched role model... To this goal I also dedicate myself".

My contributions to the University of Florida were appreciated, not only by the University itself, but also by several outside people. In particular during an official evaluation of the astronomy department of the University of Florida, Dr. F. Kerr wrote: "Another important development in faculty quality is the arrangement that has been made with Dr. George Contopoulos, of the University of Athens, a scholar of international renown in the fields of stellar and galactic dynamics. Perhaps more important is the series of workshops that he has established, in which a substantial number of scholars from this and other countries are invited to join in a concentrated study of some important area in the general field of dynamics. As a result, the department is well known in this field".

We started this series of workshops with a workshop on the "Orbits in barred and spiral galaxies" in 1985. The proceedings of this workshop were not published, but the subsequent proceedings were published by the New York Academy of Sciences, as special volumes in their Annals. The topics were the following:

2nd workshop 1986:	Chaotic Phenomena in Astrophysics, Vol.497 (1986). Editors: Buchler, J.R. and Eichhorn, H.
3rd workshop 1987:	Integrability in Dynamical Systems, Vol.536 (1988). Editors: Buchler, J.R., Ipser, J.R. and Williams, C. A.
4th workshop 1989:	Galactic Models,Vol. 596 (1989). Editors: Buchler, J.R., Gottesman, S.T., and Hunter, J.H.Jr.
5th workshop 1989:	Nonlinear Astrophysical Fluid Dynamics, Vol. 617(1990). Editors: Buchler, J.R., and Gottesman,S.T.
6th workshop 1990:	Nonlinear Problems in Relativity and Cosmology, Vol.631 (1991). Editors: Buchler, J.R., Detweisler, S.L. and Ipser, J.R.
7th workshop 1991:	Astrophysical Disks, Vol.675 (1992). Editors: Dermott, S.F., Hunter, J.E. Jr. and Wilson, R.E.
8th workshop 1993:	Stochastic Processes in Astrophysics, Vol. 706(1993). Editors: Buchler, J.R. and Kandrup, H.E.
9th workshop 1993:	Three-Dimensional Systems, Vol.751 (1995). Editors: Kandrup, H.E., Gottesman, S.T. and Ipser, J.R.
10th workshop 1994:	Waves in Astrophysics, Vol.773 (1995). Editors: Hunter, J.H.Jr. and Wilson, R.E.

In these workshops participated, besides the local people, some dozens of people from all over the United States and a number of invited participants from Europe.

The Florida workshops continued even after my departure from Florida. The 11th and 12th workshops were published in 1997 and 1998. The 13th Florida workshop was in my honor. It took place in 1998 and was published the same year. Among the participants were D. Lynden-Bell, C. Hunter, E. A. Spiegel, T. de Zeeuw, E. Athanassoula, A. M. Fridman, I. Shlosman, R. V. E. Lovelace, C. C. Lin, B. N. Miller, D. Kazanas, etc. Dr. Kazanas represented my undergraduate students, and E. Athanassoula my graduate students, who did their theses with me.

During these workshops we had many opportunities for discussions that led, later, to useful collaborations.

The 6th workshop volume was dedicated to the memory of B. Xanthopoulos, who was a speaker at this meeting, but was later killed by a distraught student of the University of Crete. Xanthopoulos was my student as an undergraduate, but he did his thesis at the University of Chicago. He was a very close collaborator of S. Chandrasekhar, who was coming often to the University of Crete, where Xanthopoulos was a professor, to collaborate with him. Xanthopoulos had become one of the leading relativists worldwide.

During a seminar at the University of Crete a student, who had complaints against Xanthopoulos, started shooting and killed him and another professor, St. Pnevmatikos, and wounded several more. The others were saved by the initiative of Dr. S. Persides (another participant of our 6th workshop) who threw a table against the intruder. Persides was wounded very seriously and barely escaped death. The distraught student later committed suicide.

My life in Florida was very active. Besides my scientific work I travelled a lot, all over the United States, giving lectures at about 50 Universities and Institutes, including a few Canadian Universities. In many places I gave lectures on several successive years.

My lectures were on recent topics and developments in dynamical astronomy. One of my lectures in Chicago was about Arnold diffusion and escapes. In the audience was Dr. Chandrasekhar. He asked me if I had a mathematical theory to describe these phenomena and I simply said: "Sorry, No". But Chandra was not disappointed. He said: "Nevertheless, this talk was the best that I have heard for a long time!"

During some of my trips I had particular difficulties. During 1993 I had planned a tour to the north, including Princeton, New York and Boston. I arrived at the Newark airport near New York and took the bus to Princeton. The weather was really beautiful. However, after my lecture I received a telephone call from Bruce Elmegreen, whom I was supposed to meet the next day in New York, telling me that the weather forecast for next day was very bad and that we should meet at a certain office of the New York University, instead of uptown New York, as we had planned. I couldn't really believe him, as the weather in Princeton was so nice.

But next day everything was covered deeply in snow, even in Princeton, and the snow was falling continuously. With some difficulty I walked with my luggage to the train station. Fortunately the train was operating. I arrived in New York, and, taking a taxi, I came to the New York University.

After some time Bruce Elmegreen arrived, but C. Yuan, who was also supposed to meet us, could not drive out of his house, and did not arrive at all. I had some very interesting discussions with Bruce Elmegreen, but the problem was how I should continue my trip to Boston. My flight was cancelled and there was no prospect for another flight the same or next day. Thus, I decided to take the train. I arrived at the station to take the first train to Boston. But the trains were very slow this day. After a few hours of delay I boarded a train to Boston. But the snow was so deep that the train would stop several times until the track was cleared. A trip that would normally last three hours took eight hours. We arrived in Boston at midnight. I was fortunate enough to take the last subway train to Harvard square. I called a friend, who lived near Harvard and was supposed to host me this night. He told me that he would come in a few minutes. But the time was passing and he did not come. I tried to call again my friend but there was no answer. Meanwhile bulldozers were clearing the streets continuously and in most places the heaps of the collected snow were higher than one meter. It was very difficult to move around. Finally my friend appeared on foot and welcomed me. He had tried to use his car but the car stuck in the snow, and finally he came on foot. Nevertheless everything was fine in the end.

On another occasion I had a lengthy trip to Calgary, Canada. I flew to Atlanta, then to Salt Lake City and finally to Calgary. There was Dr. D. Hobill, a relativist working on chaos, and I had a very nice time there. Besides our scientific discussions, he took me to Banff in the Canadian Rockies that is a well known spa in this area. I was surprised to see lakes of steaming water, surrounded by snow and people happily bathing there. Another interesting view was the deer that were walking around in the streets.

Everything was fine, but after a very good dinner, as I was going to my hotel I felt very uncomfortable. I had an unbearable pain on my back, and no aspirin was able to relieve it. It was due to my kidney. Only a hot bath could reduce the pain. I spent most of the night in the bath, and next morning I started my long trip back to Florida.

Thank God I had no serious health problems in the U.S.A. I had a health insurance, but I would not like to spend my time in a hospital. The kidney gave me two more attacks, but after that, somehow, the problem disappeared by itself.

When I went to Chicago to receive a honorary Doctor's degree (section 44) I had my wife with me. After Chicago I returned to Florida, while my wife visited a relative in West Virginia before returning to Greece. But there she became seriously ill and had to be hospitalized. She had no health insurance and my insurance would not cover her. I would have to pay several thousands of dollars, but first she had to get well. Fortunately after a few days she could get out of the hospital. When she asked how much she had to pay they said "nothing" (because my wife was a medical doctor). Thus, this difficulty was also solved happily.

Besides my wife two of my children visited me at the University of Florida. My son, John, was a graduate student at Cornell. He did his thesis under R. Lovelace on "Magnetically driven jets and winds" (1991). He found a general analytical solution of the Grad-Safranov equation that describes such jets and winds. He published several papers on this subject in the Astrophysical Journal, and he presented his first results at the 7th Florida Workshop in 1991. After his talk several people came to congratulate him. Among them was Alar Toomre, who made a remark: "I could understand the son, but I did not understand the father".

My daughter, Julie, came to Florida for one month and attended courses in English. When I looked at the list of courses offered by the University I saw, among others, courses in belly dancing I and II, astrology, aerobics, etc! Julie was at that time an undergraduate in biological sciences at the Ithaca College, in Ithaca, New York. She adapted herself easily in Florida and she became a friend with all the staff of the department.

In Florida I had the opportunity to visit several places of interest. I went to Cape Canaveral, Orlando, Disneyland, Bush Gardens, Tampa, a Greek community in Tarpon Springs, where most of the streets have Greek names, Tallahassee, and many times to the Atlantic Ocean and the Gulf of Mexico.

Many Sundays Heinz Eichhorn and his wife would take me to excursions. During these trips we had scientific and philosophical discussions that lasted for hours. Heinz Eichhorn had a solid education from his Vienna years. He often surprised me with his knowledge of ancient Greek. Our philosophical discussions were often passionate. We agreed on many things. We also disagreed on a number of points, but our discussions were always friendly. Only rarely I could discuss such problems with other American colleagues. Many times he would take me to his house for dinner and we used to hear classical music from his large collection of records.

His main interests were in problems of astrometry and kinematic astronomy. His approach was quite modern, introducing computers instead of the heavy apparatus of many coordinate transformations. We had many discussions, e.g., on problems concerning time. He often asked my opinion about his papers. He was writing a book on "Kinematic Astronomy", that I found very interesting, but he did not finish it.

During his last years he had to fight with cancer. He fought bravely to the last moment, with heavy treatment, and his wife was also fighting bravely on his side. I liked him as a person, although he was sometimes outspoken and voiced strongly his opinions. But he was frank and honest, and he did his best for the astronomy department. I will miss him.

39. INTEGRABLE MODELS

One topic that attracted particularly my interest in Florida referred to integrable models.

Integrable models are very useful in the general study of dynamical systems for two reasons: (a) Because in such cases the solutions (motions) can be found analytically, and (b) because such models can be used as starting points in the study of nearby, perturbed, systems.

The simplest integrable models are: (1) the case of two, or more, uncoupled oscillators, and (2) the Kepler problem. Many more integrable models were found by Liouville, and also by Stäckel and others at the beginning of the 20th century. These models were rediscovered in later years, and they have been used extensively recently in constructing models of galaxies (e.g. de Zeeuw, Hunter and Schwarzschild 1987). I made a systematic classification of the Stäckel models in two and three dimensions (Contopoulos 1994).

Nevertheless, these models seemed to exhaust the class of integrable models. Thus, it was a surprise when Hénon (1974) and Flashka (1974) found a very different integrable model, the Toda lattice (Toda 1970). After this discovery many more integrable models were discovered by different methods (Hietarinta 1987). The possible integrability of the Toda lattice was first found by numerical experiments (Ford et al. 1973).

The Toda lattice can be considered as a set of coupled oscillators, where the forces are exponential. In the lowest order the Toda lattice consists of a set of uncoupled oscillators, and the forces are linear. If we add higher order terms the problem becomes nonlinear. E.g. if the Hamiltonian is truncated at order 3 we find the Hénon-Heiles (1964) Hamiltonian, which is chaotic to a large degree. Hénon and Heiles state that "nothing would be fundamentally changed by the addition of higher order terms" in their model. But if one adds appropriate higher order terms one finds finally the Toda lattice, which is integrable. Thus, we decided to study successive truncations of the Toda lattice, in order to find the degree of chaos in each case (Contopoulos and Polymilis 1987).

The most surprising result of our study was that the inclusion of 4th order terms beyond the Hénon-Heiles Hamiltonian generates a system with very little chaos. This system is nonintegrable (Yoshida, Ramani and Grammaticos 1988), but appreciable chaos appears only for energies about 200 times larger than in the Hénon-Heiles case.

The addition of 5th order terms makes the system again largely chaotic, as in the Hénon-Heiles case. But as we add further higher order terms chaos is reduced. Thus, gradually, we are led to the Toda lattice Hamiltonian with no chaos at all.

Another problem connected with integrability is the search for rotating integrable models. Until a few years ago only a trivial integrable rotating model was known, namely the case of rotating homogeneous ellipsoids (Freeman 1996), which is equivalent to two uncoupled oscillators.

Thus, it was of interest that another integrable rotating model of the Stäckel type was found by Vandervoort (1979). This represents a model of a rotating bar, but with a rather special mass distribution. It can be also considered as a problem of two centers, plus a particular perturbation. We studied in detail the forms of the orbits in this potential (Contopoulos and Vandervoort 1992). The orbits are in general rings (tubes) around one or the other center, or tubes around both centers. Thus, they are very different from the orbits in the usual (nonrotating) Stäckel potentials. For this reason such a model is particularly noteworthy.

It would be of great interest to find further rotating integrable models. However, up to now, no further models of this type were found.

40. ESCAPES

A problem of interest for galactic dynamics is the problem of escapes. The usual way to find the escape rate from a cluster, or a galaxy, is described by Chandrasekhar (1942). We assume that after one relaxation time T, the system acquires a Maxwellian distribution of velocities. In such a distribution a proportion $Q = 0.0074$ of stars have velocities larger than the velocity of escape and leave the system. But after another interval T a new Maxwellian distribution is established, and a new proportion Q escapes. Thus, on the average, in a time ΔT the proportion $\Delta n/n$ of escaping stars (where $\Delta n < 0$) is equal to

$$\frac{\Delta n}{n} = -Q\frac{\Delta T}{T} \tag{40.1}$$

But this proportion is an overestimate. In fact, stars acquire large velocities by means of successive encounters with other stars. A star with a relatively large velocity, but smaller than the velocity of escape, goes to large distances for a long time and does not undergo many encounters. Hénon (1960) found that if the masses of the stars are infinitesimal, then the encounters never produce velocities larger that the escape velocity, thus we have no escapes at all. However, if the masses are finite (non zero) then some encounters near the center of the system produce large enough velocities so that some stars do escape.

Similar results are found if we consider a nonaxisymmetric model of a galaxy. In the case of a rotating galaxy, the stars beyond corotation can escape because no curve of zero velocity limits their motion outwards (if there are curves of zero velocity around the center they limit the motion inwards). However many orbits do not escape because they are restricted by another integral besides the energy, of the "third integral" type.

In a nonaxisymmetric rotating system neither the energy E nor the angular momentum J is an integral of motion. Only their combination $C = E - \Omega_s J$ (where Ω_s is the angular velocity of rotation of the system) is an integral, called the "Jacobi integral".

The energy of a star in such a system changes by encounters, especially when the star is close to the center. The energy of the star increases, or decreases, but on the average it increases. After several changes the star acquires positive energy and escapes from the system.

I presented my first results on escapes at a Symposium on the "Few body problem" in Turku, Finland, 1987 (Contopoulos 1988c). I flew from Athens to Stockholm and then to Finland in the summer of 1987. When I was leaving Athens the weather was extremely hot. My wife suggested that I should take an overcoat, but I refused. I could not believe that there could be a really cold weather in Finland. But when I arrived there I really froze to death. The weather was extremely cold. However, as we entered the meeting hall everything was warm and the friendly atmosphere and the discussions made us forget the weather.

After that meeting I was involved more seriously in the escape problem. During my stay in Florida I wrote several papers on this topic, starting in 1989 (Contopoulos 1990a). Some of these papers were done in collaboration with H. Kandrup, D. Kaufmann, C. Siopis and R. Dvorak. These studies dealt with nonrotating systems. When the energy is above the escape energy, the curve of zero velocity has one or more openings, through which a particle can escape from the system. However, the escape occurs only if the particle crosses a particular orbit across any particular opening, called Lyapunov orbit.

The Lyapunov orbits are unstable and their asymptotic manifolds reach the central part of the system. Orbits starting at points of this manifold approach asymptotically a Lyapunov orbit but never escape from the system. On the other hand, a small deviation of such an orbit outwards leads to escape. But a perturbation inwards brings the moving particle back inwards. Such a particle then may escape after a longer time. The initial conditions of escaping orbits form the so-called "escape domains" or "basins of escape".

The approach to escape is sometimes very complicated. The escape domains have a fractal structure. The asymptotic curves of some periodic orbits (section 43) make infinite rotations around the basins of escape. However, one can study the problem of escapes in a statistical way. Many interesting results were found in this way. It seems that the escapes follow certain rules that are universal, i.e. independent of the particular system considered.

The most recent paper of this series was published in the Journal "Chaos" (Kandrup, Siopis, Contopoulos and Dvorak 1999).

Another, more recent, approach was adopted by Contopoulos and Efstathiou (2004). We studied in detail the structure of the asymptotic curves and of the phase space in a simple dynamical system of 2 degrees of freedom. The particles in certain regions escape very fast from the system. In other regions they escape after one oscillation, or two oscillations, etc. The topology of these regions is very interesting. In general they form spiral sets with infinite rotations. We can see very vividly how various regions become empty, one after the other (Fig. 28b).

Although in conservative systems of finite volume there are no attractors, in systems with escapes the infinity acts as an attractor. Thus, such systems have certain properties of dissipative systems, in which attractors appear quite naturally.

41. POTENTIALS WITHOUT ESCAPES

Although in many dynamical systems there is an escape energy, beyond which we may have escapes, there are also systems without any escapes, even for arbitrarily large energies. In such a system the particles behave like the molecules of a gas inside a room. Even if their velocities may be arbitrarily large such particles cannot escape.

It has been conjectured that such systems are completely chaotic if the energy of the particles is sufficiently large. However, we found that most systems have islands of stability for arbitrarily large energies (Contopoulos et al.1994a).

In some cases the existence of islands was proven analytically. In particular we have considered the system

$$H = \frac{1}{2}(\dot{x}^2 + \dot{y}^2 + x^2 + y^2) + x^2y^2 = h \qquad (41.1)$$

for large values of the energy h. In this system the oscillations along the x- and y-axes are periodic orbits. The orbit $y = 0$ on a surface of section $(x, \dot{x})$ is represented by an invariant point $(x = \dot{x} = 0)$. This orbit is alternatively stable and unstable as h increases. Thus, beyond any given energy h there are values of h for which this orbit is stable. Therefore chaos is never complete for arbitrarily large energies.

Furthermore we have much numerical evidence, supplemented by some general theorems, showing that small islands of stability are generic in many regions of phase space.

It is of interest to see what happens to the asymptotic curves (for definitions see section 43), of the main unstable periodic orbits when the energy increases.

The lobes formed by these curves become longer and longer, while their area remains constant. But as the available space is limited and the asymptotic curves cannot intersect themselves, the lobes become very thin filaments that form spirals, both clockwise and counterclockwise, forming impressive patterns. Only the use of computers has made possible to calculate such patterns.

Some friends have suggested that it would be interesting to make an art exhibition of various patterns of asymptotic curves. In fact such patterns, together with the fractal sets of Mandelbrot (1982) are among the most nice figures that can be derived from pure mathematics.

Another important case that has islands of stability for arbitrarily large perturbations, is the standard map:

$$\begin{aligned} x' &= x + y' \\ & \qquad\qquad (mod 1) \\ y' &= y + \frac{K}{2\pi}\sin 2\pi x \end{aligned} \tag{41.2}$$

In this system we found, theoretically and numerically, islands of stability for arbitrarily large perturbations K (Dvorak et al. 2003, Contopoulos et al. 2004).

Of course these islands are very small for very large K. Therefore, if we ignore islands smaller that a limiting small size, we may state that the standard map becomes effectively chaotic if K is very large.

42. CHAOS AND RANDOMNESS

Randomness has many aspects similar to chaos, but, nevertheless, it is quite different. In fact, randomness is nondeterministic, while chaos is deterministic. If we see the distribution of the iterates of an initial point in a chaotic map, they look as being randomly scattered (Fig. 31a). But if we follow the successive iterates of a point we see that they follow a very orderly sequence in phase space (Fig. 31b). This is particularly evident if the initial point lies on an asymptotic curve of an unstable periodic orbit.

For example in the standard map (Eq. 41.2) the most simple periodic orbit is the point ($x = y = 0$). Its asymptotic curves follow a very definite pattern (Fig. 31b). For large K the directions of the asymptotic curves are close to 45°. E.g. for $K = 10$ most points of such an asymptotic curve have tangents (directions) near $\phi = 42^o$ and $\phi = 55^o$. As K increases both these values tend to $\phi = 45^o$.

On the other hand in a random system the angles ϕ are random, and not close to any particular value. Therefore, although the distribution of the chaotic points in the standard map (Fig. 31a) looks random, there is a lot of structure in this chaotic system, which is far from random.

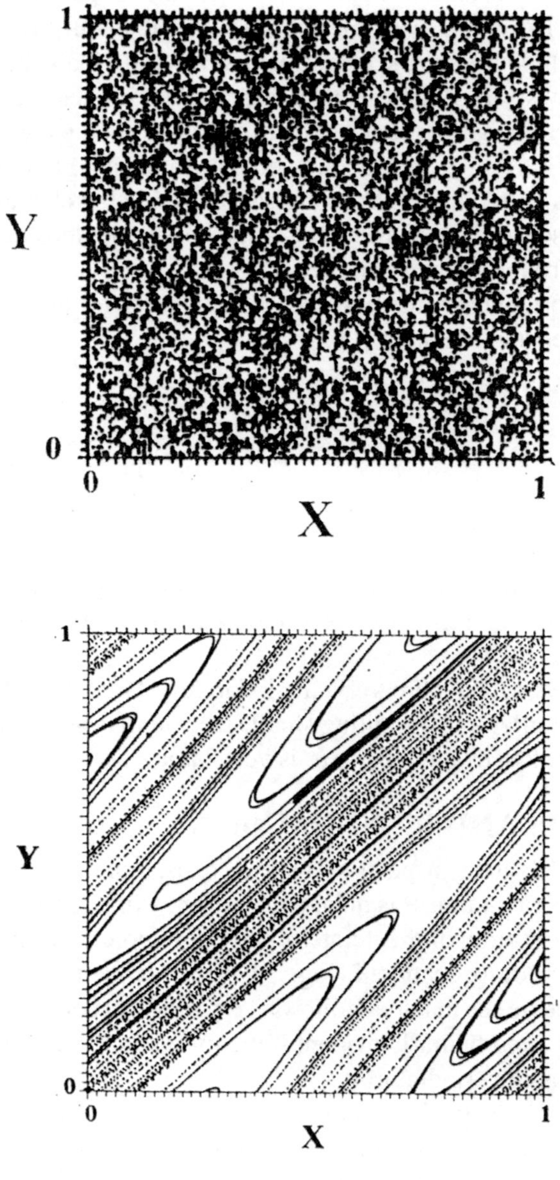

Figure 31. (a) Chaotic distribution of 10000 iterates of a single initial point. (b) The underlying order. One asymptotic curve from the periodic point (0,0).

The main reason for the non randomness of the chaotic patterns is the fact that the asymptotic curves cannot intersect themselves (i.e. the unstable asymptotic curves can only intersect stable asymptotic curves). As pointed out in the previous section the asymptotic curves take very complicated forms in order to avoid self-intersections. But this introduces another type of order. E.g. the asymptotic curves of any higher order unstable periodic orbit are contained in the gaps between the asymptotic curves of the orbit ($x = y = 0$) of Fig. 31b, therefore they are almost parallel to them.

Similar phenomena appear in generic chaotic systems. The fact that the orbits in such systems are deterministic has many consequences that differentiate such systems from random systems.

I have emphasized the differences between chaotic and random systems on various occasions (see e.g. Contopoulos 1993).

Noise is a particular aspect of randomness. The effects of noise are small, but they are basically random. Because of that the individual steps of noise are unpredictable, although their distribution follows certain statistical laws.

Some effects due to noise are the following.

Noise can generate an attractor in conservative systems, which do not have attractors. Such a system then looks as being dissipative. Noise appears also in numerical calculations of integrable systems and makes them to look as nonintegrable and chaotic. Noise increases the diffusion of chaotic orbits, or produces diffusion in cases where no diffusion is possible. Noise may also produce escapes of nonescaping orbits, etc.

The effects of noise mimic some effects of chaos, but not all of them.

On the other hand one may argue that noise always exists in Nature, therefore the noisy orbits give more reliable results than the accurate orbits in a given dynamical system. The answer to that depends on the degree of noise. If the noise is large, this statement may be true. But if the level of noise is small the real orbits follow for a long time the accurate orbits before being appreciably influenced by the noise.

In this connection there is a "shadowing theorem", which states that in a large class of chaotic systems (the so-called hyperbolic systems) a noisy orbit, starting at a given point x, is arbitrarily close to some exact orbit starting close to x (Guckenheimer and Holmes 1983). Therefore the theory of the exact orbits is a necessary background for the approximate, noisy, orbits calculated numerically.

In our numerically calculated orbits we usually check the accuracy of our calculations by using the conservation of the known integrals of motion, like the energy. These criteria are particularly useful near singular points, where the inaccuracy is most pronounced. Then it may be necessary to reduce the step of integration, or use other, more accurate, integration schemes.

It would be instructive for someone interested in problems of accuracy, to calculate orbits in the classical problem of two fixed centers (section 36). This problem is integrable, therefore there is no chaos at all. However, if a numerical orbit comes close to one of the fixed centers, the errors are significant and the orbit beyond such a point is very inaccurate. On the other hand, the exact orbit can be given analytically. Therefore, such a problem is particularly useful in checking various integration routines.

A vivid example, where it was necessary to use a very small integration step in order to find a qualitatively accurate result is the following (Contopoulos 1981c). I had found a periodic orbit in a barred galaxy, and the first calculations indicated that this orbit was very unstable. Namely, by using an integration step $\Delta t = 10^{-6}$ I found a very large stability parameter ($a = 478$), while a stable orbit should have $|a| < 1$. But by considering the invariant curves of non-periodic orbits it was evident that this orbit was stable. Thus, I repeated the calculations with steps $\Delta t = 10^{-7}$, $\Delta t = 10^{-8}$ and $\Delta t = 10^{-10}$. Only with the last step I could find an accurate stability parameter, $a = 0.46$, which was of the stable type.

The confusion between chaos and randomness has led to seriously wrong conclusions. Such an example is the book of I. Prigogine "The End of Certainty" (Prigogine 1997). Prigogine argues that trajectories are in general useless, and he proposes, instead, to use distributions of probabilities. He states: "There are two types of trajectories, 'nice' deterministic trajectories and 'random' trajectories associated with the resonances, which wander erratically".

This statement contains a number of misconceptions. First of all there are no random trajectories in deterministic systems, but chaotic trajectories, which are equally deterministic as the 'nice' trajectories (ordered orbits). The orbits near a resonance are also ordered. Only in the regions between resonances, where various resonances interact (section 13) the orbits are chaotic, and seem to wander erratically, although they are deterministic. Finally, although distributions of orbits are useful, nevertheless such distributions do not eliminate the importance of orbits.

I was in Austin, Texas, when the book of Prigogine was in press under the title "La fin des certitudes" (1996). I met Victor Szebehely, who wanted to introduce me to Ilia Prigogine. He called him, and Prigogine invited us to lunch. During this lunch he gave us a preprint of his book, and he spoke at length about its contents. I was reserved, but only when I read the book I could point out its weak points.

The views of Prigogine have been criticized severely by specialists in the field of chaos, like Per Bak (1997) in a review of the book above. He states "Chaos theory supports the classical view, not that of Prigogine. No modifications of the fundamental laws of physics are needed". His conclusion is really strong. "Every now and then, crackpot papers are submitted for publication to scientific journals. When pseudoscience is presented by a highly esteemed, Nobel prizewinning chemist, the damage is not so easy to contain".

43. HOMOCLINIC AND HETEROCLINIC TANGLES

Continuing my work on topological methods in galactic dynamics (section 22) I worked for several years on various topological aspects of chaos, like the homoclinic and heteroclinic tangles.

Chaos appears first near unstable periodic orbits. Every unstable periodic orbit in a conservative system has pairs of eigenvalues $(\lambda, 1/\lambda)$, where at least one value of λ is absolutely larger than 1. In an unstable case of a two-dimensional map we have one pair $(\lambda, 1/\lambda)$ with λ real and $|\lambda| > 1$. Along a particular direction, called eigendirection a point close to the unstable periodic point is mapped along the same direction at a distance λ times larger than the original distance from the periodic point. This is the unstable direction. Similarly there is a stable eigendirection, along which the point approaches the periodic point by a factor $1/\lambda$.

Further away from the periodic orbit the successive iterates starting along the eigendirections form four asymptotic curves, two of them unstable, starting in opposite directions (U and UU) and two stable (S and SS), again starting in opposite directions (Fig. 32).

The asymptotic curves were first introduced by Poincaré (1899). They are so complicated, that Poincaré writes: "One is impressed by the complexity of this figure, that I will not even try to draw". Nevertheless Birkhoff (1935) later, considered this problem in some detail. He introduced the notion of "signature", which is the sequence of homoclinic points along an asymptotic curve. Later this work was further developed by Smale (1963,1967), who introduced the basic chaotic structure in the homoclinic tangle, called "Smale horseshoe".

In an integrable system the asymptotic curves join another unstable point of the same periodic orbit of multiplicity m (or the same unstable point if $m = 1$) and they are called separatrices. This is because they separate invariant curves around a central point and islands around m stable points that are found between the m unstable points considered above.

But when a generic nonlinearity is introduced in the system, the unstable and stable asymptotic curves make infinite oscillations (and form an infinity of lobes), intersecting each other in a complicated way at infinite points, called homoclinic points. The set of asymptotic curves is called a homoclinic tangle (Fig. 32).

If the asymptotic curves from a given periodic orbit intersect the asymptotic curves of a different periodic orbit, these intersections are called heteroclinic points and they produce a heteroclinic tangle (Fig. 33). This type of intersections (the heteroclinic intersections) are a clear example of an interaction of resonances (section 13) therefore they are the main characteristic of chaos.

In practice, inside every homoclinic tangle there are higher order unstable periodic orbits of different multiplicities. Therefore, every homoclinic tangle is also a heteroclinic tangle with higher order unstable periodic orbits.

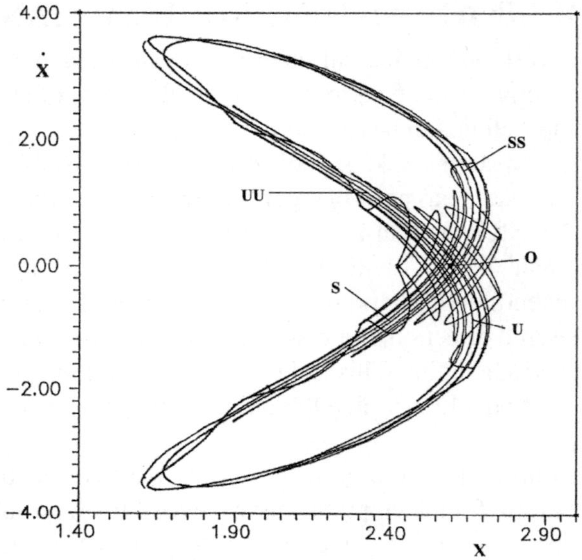

Figure 32. A homoclinic tangle, generated by the intersections of the unstable (U, UU) and stable (S, SS) asymptotic curves of the periodic orbit O.

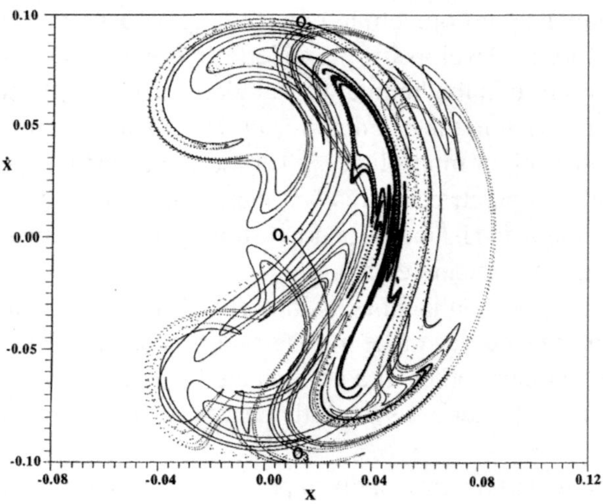

Figure 33. A heteroclinic tangle. Intersections of the asymptotic curves of a simple periodic orbit (O) and of a triple periodic orbit (O_1, O_2, O_3).

When an integrable system becomes slightly nonintegrable the homoclinic tangles occupy small domains around the asymptotic curves, mainly close to the unstable points, where the asymptotic curves make the largest oscillations. Thus the chaotic domain is limited mainly close to the unstable points. But when the nonlinearity increases, the chaotic domain increases and finally it covers most of the phase space.

Although the oscillations of the asymptotic curves are the main characteristic of chaos, they are not at all random. They follow some rules, which are not followed by random sets (section 42). Therefore the study of the homoclinic and heteroclinic tangles gives the structure of chaos in every particular case.

We studied the problems connected with the asymptotic curves in a number of papers after my first papers on resonance overlap (Contopoulos 1966d, 1971b). More recently we studied the asymptotic curves of the Lyapunov orbits that separate the escaping and nonescaping orbits (section 40) (Contopoulos, 1990, Contopoulos and Kaufmann 1992). Even more recently we made a systematic classification of the homoclinic tangles (Contopoulos and Polymilis 1993, 1996; Contopoulos et al. 1996; Contopoulos and Dokoumetzidis 2004). Our study was based on accurate numerical integrations, while previous authors usually made sketches that sometimes missed some of the properties of the homoclinic tangle. In this connection Lynden-Bell (1998) wrote: "in Saltjobaden in a conference on the dynamics of barred galaxies, George ... taught us how to measure and classify chaos, even complete chaos!".

There are further useful applications of the homoclinic tangles. Such an application was in finding the minimum Poincaré recurrence time (Contopoulos and Polymilis 1996).

Poincaré formulated his famous "recurrence theorem" in 1899. This theorem states that an orbit will return in general arbitrarily close to its initial position an infinite number of times. The only condition is that the phase space is finite. Then a set covering an area of measure m is mapped at every iteration to a different set of measure m (because of the area preservation). But if the total area is finite (M) there cannot be only nonintersecting sets after N iterations if $N > M/m$. Thus, there will be recurrence and the (average) Poincaré recurrence time is $N = M/m$.

However, the recurrence time is different in different domains of phase space. The question is what is the minimum recurrence time. This minimum time cannot be arbitrarily small. But we found cases where the minimum recurrence time is 200 times smaller than the average Poincaré recurrence time. Furthermore the minimum recurrence time can be defined also in cases with escapes, where the phase space is unbounded (infinite) and no average recurrence time can be defined.

44. HONORARY DEGREE FROM THE UNIVERSITY OF CHICAGO

An unexpected distinction from the USA was a honorary doctor's degree from the University of Chicago, that was awarded to me on the occasion of the 100th anniversary of the University, in 1991.

At the same time a number of distinguished scientists of various disciplines (10 in sciences, 8 in letters and 2 in law) received the same degree. A second astronomer, Barbara Burbidge was also among the recipients. My wife was also invited. So we came to Chicago around the first of October 1991.

The celebration was very formal, in the large chapel of the University of Chicago. All the recipients of honorary degrees were wearing official robes, provided by the University of Chicago. They were invited, one by one, by the rector of the University, Mrs. Hanna Gray, who read the citations and then gave the hood of the University of Chicago (Fig. 34). In my case the citation read: "You have discovered an unexpected richness and variety in the properties of stellar orbits in galaxies, and you have shown that such phenomena must manifest themselves in new and distinctive ways in the structures and internal motions of galaxies. In your own work and in your enthusiastic and generous encouragement of the work of younger colleagues, you have laid the foundation and created the framework for major advances in galactic dynamics".

Figure 34. Award of a honorary doctor's degree by the University of Chicago (1991).

When we were entering the hall, I was introduced to some recipients of honorary degrees. One of them asked: "Are you a Greek-American?" "No, I replied, I am a Greek-Greek".

After some music we paraded, lead by the philharmonic band, through the streets, to the main building of the University. The weather was excellent. It was fortunate, because the previous evening there was a terrible rain.

During the official dinner there were several informal speeches. The rector noted in her speech that Dr. Contopoulos had his birthday the same day of the inauguration of the University. She noted that the probability that any one of the recipients should have this chance was less than 5%.

Another (unofficial) speaker remembered that some years ago the queen of England visited Chicago, and she wanted to be awarded a honorary doctor's degree. Her wish was transmitted to the University in a discrete way by the diplomatic channels, but when the University had to decide, a member of the Senate asked: "What scientific achievement was done by the queen of England"? This remark killed the proposal, and no degree was awarded.

When we returned to our hotel we found a huge pack of flowers sent to us by the rector. In general my wife and I had a very good time in Chicago.

During my visit to Chicago I gave a lecture and I met several old friends.

On this occasion, I was also invited by Dimitri Mihalas to give a lecture at the University of Illinois in Urbana. The day I was going there the whether was very bad. A terrible storm was expected. Nevertheless our plane flew with some delay to Urbana and my talk took place as scheduled. But then the storm struck. Dimitri Mihalas took me to the airport and when the first rain subsided the departure of the flight was announced. We went to the plane, but before entering it we were turned back. The storm was getting worse. After some hours, they told us that they would drive us to Chicago by bus. We started while the rain was at its highest. I wondered how the driver could see his way. Halfway to Chicago the bus broke down. We could only stop at a small shop on the side of the road. After some more hours another bus came in the rain to take us. We arrived in Chicago well after midnight.

I take this opportunity to say a few words about Dimitri Mihalas. His work on stellar atmospheres is well known and it is widely appreciated. Dimitri is Greek, but of second generation. His Greek language is rather poor. When he went to his native town in Crete, he entered a shop to ask directions about the road. "Who are you?" asked the shop-keeper. Dimitri said: "I am just an American tourist". "What nonsense is that?" replied the shop-keeper. "Are you not the son of M. Mihalas, who died some years ago"? The face of Dimitri resembled so much to his father's that he could not escape notice.

45. DYNAMICAL SPECTRA

Starting in 1966 I worked for several years on chaos in dynamical systems. My 1966 paper (Contopoulos 1966d) was one of the two first papers to calculate the transition from order to chaos (section 13).

Later I studied in detail chaos in galactic systems (e.g. Contopoulos 1967c, 1983b). Since 1978 I used extensively the Lyapunov characteristic numbers (section 28) in order to distinguish between ordered and chaotic orbits (Contopoulos, Galgani and Giorgilli 1978). The Lyapunov characteristic nymber (LCN) is the average exponent of the divergence of nearby orbits in a dynamical system. As explained in section 28 two particles starting at an infinitesimal distance ξ_0 at time $t = 0$ have a deviation ξ at time t. Then the average rate of exponential deviation in time t is

$$\chi = \frac{\ln |\xi/\xi_0|}{t} \tag{45.1}$$

and the limit of χ when t goes to infinity is the Lyapunov characteristic number (Eq. 28.1).

In order to see numerically the convergence of χ we need usually a large number of periods. In several galactic models we had to calculate orbits for one million periods. But the age of a galaxy (Hubble time) is no more than 100-200 periods. Then what is the use of quantities defined over millions of periods?

In 1988 I was refereeing a paper by Udry and Pfenniger (1988) on order and chaos in galaxies. They considered average exponential deviations of orbits over a Hubble time and they called that "Lyapunov characteristic number". I suggested to them to introduce a new term "Lyapunov characteristic number over a Hubble time", but they preferred not to do so. That was a mistake, because later, the term "Lyapunov characteristic number over a short time" became very popular.

Later Froeschlé at al. (1993) introduced the shortest possible interval for defining a "short-time LCN" namely one iteration in the case of a map. They presented their results at the 1992 Humboldt Symposium in Ramsau, Austria. They introduced a spectrum of the "short-time LCNs", i.e. the distribution of such "short-time LCNs". The lecture of Froeschlé looked interesting to me, but I did not pay further attention to it.

When I came back home, I found my associate Voglis working on a fast method to distinguish between ordered and chaotic orbits. This was an interesting subject and I got involved seriously in it. Voglis was using the shortest possible LCN, that he called the "stretching number". It is given by the formula (45.1), when t=1 iteration. The new result was that the spectrum of stretching numbers is invariant in a chaotic domain, while it is changing in general in an ordered domain. The usual LCN is the average value of the spectrum. It is positive for chaotic orbits, while it is zero for ordered orbits.

We prepared a paper on this subject (Voglis and Contopoulos 1994), but after it was accepted we noticed the published paper of Froeschlé et al. (1993) and we added a reference to it. Later we realized that several other people had used already "short time Lyapunov characteristic numbers". However our discovery that the spectrum of such numbers is invariant in chaotic domains was new.

When I was speaking about this problem at a seminar in Columbia University, someone in the audience claimed that such an invariance was obvious, in view of the ergodic theorem. But K. Prendergast who was organizing this seminar, remarked that this is not true. In fact the values of stretching numbers are not smooth functions of the coordinates. On the contrary the domains with given values of stretching numbers are fractal and near each point there are points with very different values of stretching numbers (Contopoulos et al. 1995a,b) One needs an extension of the known ergodic theory to fractal sets to deal with such phenomena. Such a theory does not yet exist.

Besides the stretching numbers, it is interesting to find the distribution of the angles formed by the vectors joining points on nearby orbits. We defined the helicity angles (angles between these vectors and a fixed direction, like the x-axis), and the twist angles, i.e angles between successive vectors. Correspondingly we have spectra of helicity angles, and spectra of twist angles, which are also invariant.

The use of the spectra allows us to distinguish very fast between ordered and chaotic domains. In fact we can separate ordered and chaotic domains along a particular line in phase space by calculating orbits at successive points along this line for only 10 periods (after the first 10 transient values, i.e. a total of 20 values) (Contopoulos and Voglis 1997), while other methods require at least some hundreds or thousands of values.

The dynamical spectra (spectra of stretching numbers, helicity angles, or twist angles) have proved to be a source of very useful information in the study of dynamical systems. Several papers on this subject were published, in collaboration with Drs. Voglis, Efthymiopoulos, Skokos, et al. Similar results were published in recent years by Froeschlé and his associates in Nice, and by Dvorak and his associates in Vienna.

An extension of this work to the case of Hamiltonian systems was made with Haywood Smith of the University of Florida (Smith and Contopoulos 1996). In this case the shortest time over which we can calculate a stretching number is the integration step itself. The information provided by using such a small step is sufficient to explain the spectra that we find with any larger interval of time Δt.

46. DESTRUCTION OF ISLANDS OF STABILITY

For several years I spent a lot of effort in trying to understand the qualitative characteristics of generic dynamical systems. For this purpose I did extensive numerical investigations. I had noticed, starting in 1960, that some simple systems have the main qualitative characteristics of more complicated realistic models. The same remark was made also by many other authors. But this statement must not be overstressed. Sometimes there are exceptions. E.g. by adding appropriate higher order terms to the Hénon-Heiles (1964) model, which is very chaotic for large energies, one finds the Toda (1970) lattice, which has no chaos at all (section 39).

A problem of special interest is the increase of chaos in a strongly nonlinear dynamical system, which leads to the destruction of the islands of stability.

A system that is close to an integrable one has very little chaos, mainly around its unstable periodic orbits. In general such a system has many islands of stability of various orders (sections 6 and 22). Between these islands of stability there are unstable periodic orbits surrounded by chaotic domains. As the perturbation increases the islands become larger, but the chaotic domains become also larger. Finally beyond a critical perturbation the chaotic domains of various resonant types interact and we have a large degree of chaos. This is the phenomenon of resonance overlap, that we discussed in section 13.

As the perturbation increases further, the islands of stability become smaller and finally they disappear. The question is how the various islands are destroyed.

A systematic study of this phenomenon was made in 1999 (Contopoulos et al. 1999b). Every island of stability is limited by a last invariant curve (last KAM curve), beyond which there is a large chaotic domain, the so-called "large chaotic sea". The island is composed of an infinity of closed invariant curves, but between them there are small chaotic zones, around unstable periodic orbits corresponding to higher order resonances.

As the perturbation increases these higher order chaotic zones increase in size. Close to the outermost invariant curve (the "last KAM curve") there are several resonances, but usually most conspicuous is a relatively low order resonance, which generates a relatively large chaotic domain inside the last KAM curve. Then, at a critical perturbation the last KAM curve (KAM torus in the 3-Dimensional phase space) is destroyed by developing an infinity of holes, and then it is called a cantorus, meaning a torus with infinite holes, forming a Cantor set. Then the size of the island decreases abruptly. The island is then limited by a new "last KAM curve", inside the resonant chaotic region, and this resonant region now communicates with the outer large chaotic sea.

As the perturbation increases even further a new chaotic domain is developed inside the new "last KAM curve" and beyond another critical perturbation the new "last KAM curve" is again destroyed and the island decreases once more abruptly in size.

Between these abrupt decreases the size of the island increases as the perturbation increases. Therefore the function giving the size of the island as a function of the perturbation has a very strange form. It has infinite abrupt decreases, but also regions with an overall tendency to increase. As the perturbation increases substantially the overall size of the island decreases and tends to zero. This happens as follows. After the formation of higher order resonances of the form $1/m$, where m increases and tends to infinity, the periodic orbit at the center of the island becomes unstable at a critical perturbation. For larger perturbations a chaotic domain is formed around the center, but it is still surrounded by a set of closed KAM curves. For an even larger perturbation the "last KAM curve" is destroyed and the chaotic domain close to the center communicates with the outer chaotic sea. Then, the original island does not exist anymore.

However, this is not quite the end of the existence of the islands. In fact, when the orbit at the center of the island becomes unstable we see the formation of a double period stable periodic orbit, or of two stable periodic orbits of equal period. Thus, when the original island is destroyed, there are still two "daughter" islands around it.

These daughter islands are destroyed by a similar process and four grand-daughter islands are formed, and so on. The final destruction of all the off-spring of the original island occurs within a rather small interval of increasing perturbations. In fact we know that each successive generation of islands lasts about $\delta = 8.72$ times less (in the parameter range) than the previous generation (section 29). Therefore, after the destruction of the original island the end of the whole set of secondary islands appears rather fast, i.e. after a small further increase of the perturbation.

Does this mean that for even larger perturbation chaos prevails everywhere and no order exists anymore? No. In fact, at certain large perturbations new families of periodic orbits appear, that have no relation with the previous families. These are the "irregular" families of periodic orbits (section 22), which contain also stable periodic orbits. Stable periodic orbits are surrounded by islands of stability. The evolution of these islands is the same, as in the cases of regular periodic orbits. Namely, as the perturbation increases the islands increase in size and later they decrease to zero after infinite increases and decreases.

Such irregular periodic orbits and islands were found for arbitrarily large perturbations in some Hamiltonian systems (e.g. the systems (5.1) and (41.1)) and in the standard map (Eq. 41.2). Therefore, it seems that chaos is not complete, even for arbitrarily large perturbations. However, in practically all cases of large perturbations the size of the islands of stability is very small. Therefore we may say that in such systems chaos is "practically" complete.

The appearance and disappearance of islands of stability in more degrees of freedom is a largely unexplored subject. There is no doubt that such studies would be highly interesting for a better understanding of dynamical systems in general.

47. STICKINESS

Many years ago (1970) I tried to find numerically the last KAM curve around an island of stability on a Poincaré surface of section. I calculated numerically several closed invariant curves (KAM curves) around the center of the island. The invariant curves were defined by the successive iterations of an orbit by the surface of section. I was taking initial conditions x along a line starting at the center of an island x_0, with x varying by a given step Δx. Suddenly after a certain distance x the iterates of an orbit were spread out in a chaotic way. Thus, I tried to find the last closed curve by using smaller and smaller steps Δx. I noticed something strange. I found a few orbits that seemed to form closed invariant curves for a few hundreds of iterations, but later their intersections by the surface of section were spread chaotically. At first I thought that this was a numerical effect. Thus, I calculated the orbits with greater accuracy but I did not find any difference. Then I tried a very different check. I took the last point of an orbit as initial condition and calculated again the orbit backwards in time. In this way I saw the successive intersections in the opposite sequence. Initially the points were scattered in the chaotic domain but later they came close to the island and they started moving in a regular way around it, as if they were defining an invariant curve.

This was the first example of a "stickiness" phenomenon (Contopoulos 1971b (Fig. 35); Shirts and Reinhardt 1982; Karney 1983). Later this phenomenon was observed by many authors. In fact many people have found orbits that are sticky for a long time but then they go out in the chaotic domain.

This phenomenon is now explained as follows. An island of stability is limited by a set of KAM curves. When the perturbation ε increases a secondary chaotic zone is produced inside the last KAM curve, but this does not communicate with the outer chaotic domain until ε reaches a critical value ε_{crit}. For ε beyond ε_{crit} the last KAM curve is destroyed, i.e. it develops an infinity of holes, and the inner chaos communicates with the outer chaos. This "curve with holes" is an invariant set, which is called a "cantorus" (section 46). For ε slightly above ε_{crit} the holes of the cantorus are small and the communication of the inner with the outer chaos is quite slow. Thus, an orbit starting in the inner chaotic zone remains there for a long time, and produces the stickiness effect, before going out to the large chaotic domain. For even larger ε the holes are larger and the stickiness is less conspicuous.

A detailed discussion of the stickiness phenomenon in connection with cantori has been given by Efthymiopoulos, Contopoulos, Voglis and Dvorak (1997).

On a surface of section a sticky orbit is represented by a gray region formed by a large number of points, scattered nearly uniformly in a large chaotic domain, plus some darker regions containing the sticky points of the same orbit. The question is now how important are these sticky domains. It is to be noted that an orbit may enter a sticky domain several times, but most of the time it remains out

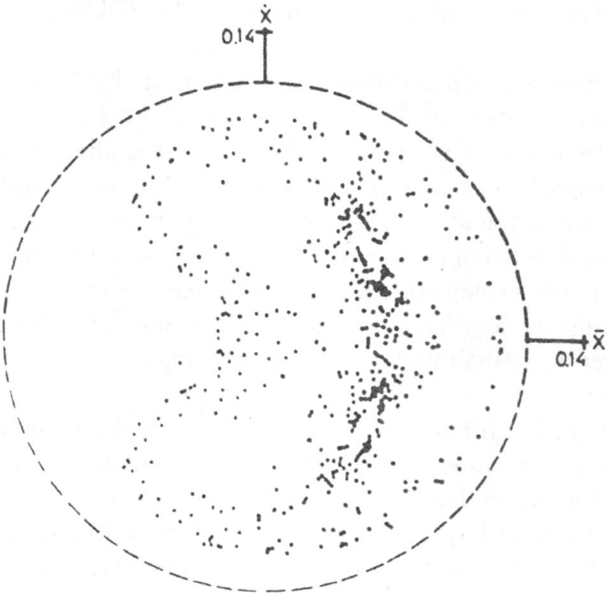

Figure 35. The first example of stickiness around two islands surrounded by a large chaotic domain.

of the sticky domains. Thus, after a very long time the darkness of the nonsticky chaotic domain will become equal to the darkness of the sticky domain. Some evidence for that effect was provided by Contopoulos et al. (1995c). But this sticky domain is subdivided by many (infinite) co-existing cantori. The innermost layers are more sticky in the sense that the orbits require a much longer time to get out of it. This time seems to tend to infinity exponentially as we approach the inner limit of the sticky zone, which is the last KAM curve around an island of stability (Contopoulos et al. 1997). Thus certain people claim that no equipartition of points between the sticky and the nonsticky domains can be expected. If this is so the cantori would allow a density difference between the sticky and nonsticky domains to persist for ever.

This phenomenon is similar to a Maxwell demon (Zaslavsky 1995, 1999) that preserves a temperature difference between two rooms that communicate through a hole. Namely, if a demon allows only fast molecules to go to the right and only slow molecules to go to the left, then the room on the right will preserve a higher temperature than the room on the left.

But this effect requires the action of an eye, or of an apparatus, that "sees" the molecules that approach the hole, and decides which are to pass through and which must be turned back. On the contrary in the stickiness case the effect is automatic.

Whether there is an automatic Maxwell demon in sticky dynamical systems depends on certain factors.

(1) On the extent of the various sticky domains surrounding an island and on the average stickiness time in every sticky domain. In many cases the sticky domains are so small that they can not produce appreciable effects.

(2) On the range of parameters in which such stickiness effects are observed. We have noticed that a small change of the nonlinearity parameter may change the stickiness time by very large factors. It is our experience that the holes of a cantorus increase considerably in size after a relatively small increase of the nonlinearity parameter.

(3) On the original distribution of the orbits. If the density of initial conditions inside a cantorus is the same as outside it, there is no reason to observe later a pronounced difference of density.

Therefore it is rather improbable to find an effective Maxwell demon in a dynamical system. Nevertheless this subject is worth of further study and much numerical work is presently in progress in this area.

48. COLLABORATORS IN GREECE

I was happy to find several good collaborators in Greece. Most of my students and assistants, that did their PhD with me (over 40) proved to be very good scientists and had a remarkable career. Among the persons that did their PhD or docent's thesis with me were: G. Antonakopoulos, E. Athanassoula, B. Barbanis, G. Bozis, E. Chaliassos, D. Dionysiou, P. Grosbøl, J. Hadjidemetriou, N. Hiotelis, P. Laskarides, E. Livaniou, C. Mertzanides, P. Michaelidis, P. Niarchos, T. Papayannopoulos, P.A. Patsis, C. Polymilis, J. Seimenis, N. Spyrou, C. Terzides, H. Varvoglis, L. Zachilas, M. Zikides, etc. Most of them became later professors of various levels (assistant, associate, or full professors).

As examples I may mention the following: J. Hadjidemetriou, professor of mechanics at the University of Thessaloniki and president of Commission 7 (celestial mechanics) of the International Astronomical Union. He is considered as one of the leading persons in celestial mechanics worldwide. B. Barbanis, professor of astronomy at the University of Patras, and later my successor at the University of Thessaloniki. He is well known for his work in galactic dynamics. G. Bozis, professor of mechanics at the University of Thessaloniki; his most important work is on the inverse problem of celestial mechanics. E. Athanassoula, astronomer at the Observatoire de Marseilles; well known for her work on galactic structure and dynamics. P. Grosbøl, who did his thesis with me on the birthplaces of nearby early-type stars. He became director of the Image Processing Section in ESO and he made interesting contributions in observational galactic dynamics. P. Laskarides , who became my successor as professor of astronomy at the University of Athens. N. Voglis, associate professor at the University of Athens and, later, director of the astronomy center of the Academy of Athens. He has made significant work on dynamical spectra, N-body simulations, and (recently) solitons and breathers in galactic dynamics. P. A. Patsis, research associate B (equivalent to associate professor), well known for his work on galactic dynamics. C. Polymilis, assistant professor, with many contributions in the theory of chaos. Unfortunately Dr. Polymilis passed away in 2000.

From the younger generation I should mention Drs. C. Efthymiopoulos, C. Skokos, H. Papadaki, E. Grousousakou and M. Harsoula. Some even younger assistants still work on their PhD theses.

Some of my undergraduate students became later well known worldwide, like Drs. B. Xanthopoulos (a leading relativist) and D. Kazanas (first to propose the inflationary cosmological model).

Only in a few cases I had difficulties with some of my students. I remember that two of them came one day to complain that I was asking them to do a lot of calculations, but their theses were not progressing. I explained that all these calculations should be parts of their theses. One of them was convinced, and in one year he finished his degree. The other said that he preferred to go to Germany and do his thesis there. But after two years he came back to me and said that there was no prospect to do a new thesis there, so he asked if he could continue his work on density waves that he had started with me. After one more year he completed his thesis that was submitted and accepted in Germany (Bonn).

I supported several of my students and assistants to get scholarships and go abroad. Most of them (but not all) did very good work there and left excellent impressions wherever they went. Among them were Drs. Barbanis, Moutsoulas, Hadjidemetriou, Bozis, Varvoglis, Polymilis, Pinotsis, Hiotelis, Patsis, Efthymiopoulos, Skokos, etc.

Several of my students continued their collaboration with me after their degree. Thus, I continue to have collaborations with the Universities of Thessaloniki, Patras, Crete and the Aegean.

I had also collaborations with colleagues of other disciplines. The most noteworthy case was that of Dr. S. Farantos, professor of physical chemistry at the University of Crete, with whom we made some interesting comparisons between classical and quantum mechanics (section 37).

During my trips abroad I often met some of my students when I was giving lectures in various Institutes and Universities. In some cases these meetings were rather unexpected. I will mention only two such cases. In the first case I was flying to Rochester, New York, during one of my first visits to the United States. The plane stopped at Syracuse, but then it broke down. It was rather late in the evening and we had to stay overnight in Syracuse waiting for another plane for Rochester. Then I saw a small plane boarding for Ithaca, New York. Ithaca is rather close to Syracuse and Rochester. I remembered that I had a student in Cornell University (I had his address but no telephone number). Thus I decided to go to Ithaca to meet him. I reached Ithaca and I soon found the apartment of my student, but he was not there. A neighbour told me that he was probably in the Cornell library at that time, up the hill from the city. I went there and I did find him, deeply absorbed over his books. You can imagine his surprise when he saw me leaning above his head.

The second case was even more unexpected. I was in Tenerife, in the Canary Islands, to give some lectures at the Instituto de Astrofisica and to visit the Observatory on the Teide mountain of Tenerife. Then my Spanish friends suggested that I should, by all means, visit the big Roque de los Muchachos Observatory in La Palma island. I flew there and a driver was expecting me to take me to the Observatory. We had a long drive up the mountain and we reached the site of the Observatory at 2200 m, near a large caldera of an old volcano. This Observatory includes several large telescopes from various countries. After our first tour around the site we entered the cafeteria. And then I was very surprised to hear a girl shouting in Greek: "Mr. Contopoulos". She was a former student of mine, connected with the Cambridge Observatory in England, and she was at that time in La Palma observing at the British telescope there. It was a real surprise for me.

My students were kind enough to organize in 1998, a Symposium in my honour, on Order and Chaos, in Livadia, Greece. Dr. J. Hadjidemetriou, one of my first students, gave the Laudation lecture. I was really moved by this token of appreciation from my former students.

49. COLLABORATORS ABROAD

A number of people did their theses abroad under my supervision (either completely, or partly). Such were the cases of P. Magnenat (Geneva; one of the first to work on systems of three degrees of freedom), J. Colin (Besançon; 3éme cycle thesis on the corotation resonance in galaxies), E. Mahon (University of Florida; prograde orbits in 3-axial galaxies), D. Kaufmann (University of Florida; self-consistent models of barred galaxies), C. Siopis (University of Florida; self-consistent models of elliptical galaxies), and Y. Papaphilippou (Paris; application of Laskar's frequency analysis method to galaxies).

I was also in the thesis committee of a number of people abroad, like C. Froeschlé (Nice), and J. Donner (Cambridge).

My collaborations with colleagues abroad resulted in joint papers with many people in several countries, like S. Chandrasekhar (University of Chicago and Yerkes Observatory), L. Woltjer (Yerkes Observatory), B. Strömgren (Princeton, Institute for Advanced Study), B. Bok (Tuscon), S. McCuskey (Cleveland), P. Grosbøl (ESO), A. Giorgilli and L. Galgani (Milan), P. Magnenat and L. Martinet (Geneva), S. Gottesman, J. Hunter and M. England (University of Florida), P. Vandervoort (Chicago), H. Kandrup, D. Kaufmann and C. Siopis (University of Florida), B. Grammaticos and A. Ramani (Paris), H. Smith (University of Florida), R. Dvorak and his collaborators (Vienna), C. Froeschlé and his collaborators (Nice), L. Ossipkov (St. Petersburg), etc.

But I had also useful discussions with many more people that are reflected in over 300 acknowledgements in their papers. I had opportunities to see many of these people at various meetings, or whenever I was invited to give lectures abroad.

Some of these collaborations were formalized as projects, which were funded by various agencies. Among them I should mention a joint project with some colleagues at the University of Florida (S. Gottesman and J. Hunter), and a "Human Capital of Mobility" project of the European Communities, that included people from Athens and Thessaloniki (N. Voglis, J. Hadjidemetriou and H. Varvoglis), Milan (A. Giorgilli and L. Galgani), Nice (C. Froeschlé, E. Lega and R. Gonczi), Vienna (R. Dvorak and E. Lohinger), ESO (P. Grosbøl), and Namur (J. Henrard). As a result of this collaboration we published a paper on transition spectra of dynamical systems (Contopoulos et al. 1997). Similar collaborations continue even today.

I supported many people from several countries, by writing hundreds of recommendation letters. It seems that in many cases such letters were successful. In one case I had written a letter supporting a graduate student from Florida to find a position at the University of Boston. When I gave, later, a lecture at this University, the Chairman came to thank me for my earlier recommendation. The student that I had recommended proved to be exceptionally good. I had a similar experience in the universities: Columbia, Florida, Florida State, Virginia, Chicago, Manchester, Vienna, Milan, Leiden, etc.

But I was not always successful. I had written a recommendation letter for a project of an American colleague, which was not approved. Some time later I happened to meet the head of the NSF department that was responsible for these projects, and I asked him why such a good project was not funded. His reply was "but your own support was not very warm". I protested strongly. I said that my recommendation was among the strongest I had ever written. But it seems that funding agencies are used in recommendations with the strongest possible words, plenty of superlatives, etc. On the other hand my style is different. I always make a balanced report, stating the good, but also the weak points of a project, or of a person, and I very rarely use words like "excellent", "very original" etc. Thus, my judgements should be taken at face value.

I remember the case of a student in California, for whom I had written a good, but not excellent, report. This student called me on the phone and asked me to write that he was my best student, etc. I refused to do so. Then the Chairman of his department called me to ask further information. I gave a moderate reply. I was not able to say anything better, and I am sorry that this student was disappointed.

The success of the students that I recommended opened the door for further students. That was, in particular, the case with the University of Manchester, that supported several students of mine.

But in another case a student in another department quarrelled with the Chairman and left the department. Then he came to me and asked me to write another letter of recommendation for him. I refused to do so. I told him that, while others had made his own entry easy, he had closed a door not only for him but for any further student in that department.

This case was exceptional. In general my students were very conscientious and gave a very good impression. Whenever I go abroad I often find former students of mine that work there. Many of my students and collaborators had excellent careers abroad, or in Greece, and I am proud of them.

50. OUR FACULTY

After my appointment as professor in 1957 I made many efforts to improve the standards of the professors in our faculty. I tried hard to overcome the influence of politics and of various cliques among my colleagues. I never tried to form a clique of my own. Even if a person that I supported was elected, I never tried to attach him to me in any way. Because of that I did not make any compromises, and sometimes I had to disagree with some of my friends.

I remember a case of a professorship in physics. My friends in the faculty supported one candidate, while I supported a different person. They told me that my candidate would generate problems to us. I replied that only the scientific level should be considered. Unfortunately I was wrong. My candidate was elected, but later, he proved in many cases to be politically and not scientifically motivated.

I had a similar disappointment later with a candidate at the University of Athens. They warned me that this person was a militant communist. I replied that I was only interested in his scientific work. My support was instrumental for his election. But after his appointment he became the leader of a faction in our faculty. I was surprised by the attitude of some people, who were following him, despite their own beliefs. I remember vividly one such occasion. A colleague sitting next to me was strongly opposed to the appointment of some assistant professors that had not been elected, but were only hired. While he was speaking my former protégé blinked his eye in a characteristic way. The speaker stopped abruptly. A little later he went to speak privately to him and when the time of a vote came, he voted against his original opinion. I learned later why. In this category of people there was a party member that should be supported by all means. In one of my confrontations with my former protégé, I asked him: "Why did you vote for Mr. so and so! You, yourself, told me some time ago that he was not worthy of promotion". My colleague sighed and said: "Oh, Mr. Contopoulos, you cannot understand. You have many more degrees of freedom than I have!" It was a really painful confession. He was a "leader", but he was not free to decide. The decisions came from other centers of power.

I was often surprised why I failed several times in our faculty, while I had succeeded in some international bodies, like the Board of Directors of Astronomy and Astrophysics. In all cases I used the same method. I was well prepared for every meeting, read carefully the related documents, and had some solutions to propose. This method worked well in unbiased environments. But when I had to fight against political interests, no arguments were of any use.

My last comments do not imply that all international bodies are exempt of bias. I had many unpleasant experiences of purely political nature when I was at the NATO Science Committee, as I will describe in section 51.

During the Junta period in my country (1967-74) many people lost their positions, because they were accused by their enemies. Their dismissal was usually followed by the following words. "This decision cannot be appealed at a court of justice". Thus, the poor people had no right to defend themselves.

During one massive expulsion of 1/3 of all professors, a number of professors, including me, appealed to the Supreme Court. Our case was so clear, that the Supreme Court members that we approached assured us that the decision of the Court would be in our favour. But the decision was against us. The day before the decision, the Court members got threatening telephone calls from the Junta, and they changed their decision.

After the Junta was overthrown several people that had strong ties with the Junta were fired. There was a "de-juntization" committee for the Universities that examined all cases. This committee was formed while I was out of the country. When I came back the general secretary of the ministry of education, who was a friend of mine, called me and said: "Where have you been these days? I wanted to put you in this committee and I could not find you anywhere".

Nevertheless I did handle two cases myself. One of our colleagues was accused by some militant students of being a Junta supporter. I happened to know this colleague, and he always attacked the Junta during the previous years. I called the students and I asked them: "Why did you accuse this professor?" And I explained his behaviour during the Junta period. They said: "The accusation was done by Mrs X, an assistant of the professor". I asked them: "Can you give me evidence of that"? Then the students visited this assistant, presumably for further information. But one had a tape recorder in his pocket. When we heard the tape recorder it was plain what happened. The assistant blamed her professor but repeated again and again: "Do not say to any one that I gave you this information".

At that time all the positions of the assistants had to be reconsidered by the faculty. Only if the vote of the faculty was positive, the assistants would continue in their positions. In the case of this particular assistant, I explained to the faculty that she had wrongly accused her professor. Many people could not believe me. Then I said: "Wait a minute", and I let them hear the tape recorder. It was like a shock.

On the other hand there were people that had acted so unjustly against their colleagues during the Junta period that they should be punished by all means. One such person was the former rector of the University of Thessaloniki (not elected, but appointed by the Junta). He was responsible for the firing of many professors of this University. In the second Junta he even became minister of education. His fame as a person with many political ties everywhere was so strong that no one dared to depose as a witness in his case.

The general secretary of the ministry had told me that this former rector would be fired because the new law stipulated that those who had been ministers during the Junta period would be fired automatically. But next day the law was published and the word "automatically" had been replaced by "if judged guilty". The new rector of our University called me and told me that there was a risk that the former rector would be acquitted for lack of evidence. He asked me if I knew someone that could go as a witness. I said that I would go myself, and I did go. The former rector was finally fired.

After the "de-juntization" period there was a strong anti-establishment movement among the students that lasted a number of years. It was like the French movement of May 1968, but it lasted much longer. There were just demands, but also some quite unreasonable requests. One extreme case was that some of my students wanted to approve the examination subjects before every written examination.

When I was elected professor at the University of Athens (1975) I found the students striking against the interim professor of astronomy, because he had, presumably, asked some questions outside the approved course material. I saw these questions and they were rather elementary. So I sided with the professor and the students turned against me. I had to give another written examination, but the striking students did not allow any one to enter the examination hall. I tried to argue with the strikers, but they started to shout against me. I told them: "If you claim that you have fought against the Junta, let us compare what you and I have done". Then a student who was probably a leader turned to his colleagues and said: "Shut up". Then he turned to me. "We know, Sir, what you have done, and we respect it. But in the present case we have to fight for our rights and we will do that to the end". The end was not so brilliant. The faculty appointed another professor to repeat the examinations, and he passed most of the students. So there were no students left out to strike.

In later years I had sometimes to face more serious problems. In one case we had in our faculty to consider the promotion of some assistant professors. During the first session I was negative in some cases. Then, I received an anonymous letter threatening to attack my children if I did not change my attitude in the following cases. I presented this letter at the faculty meeting and I demanded to stop all elections until the state would provide bodyguards for us and our children. I argued that this was done already in the case of judges in criminal courts[11]. But the majority felt that the elections should not be postponed. Then I walked out of the faculty meeting (and my absence counted as a negative vote). But my action was not an easy one. Another professor who received a similar letter changed his attitude from negative to positive.

Fortunately the situation has improved considerably the last several years.

51. NATO

After the re-establishment of a normal political situation in Greece (1975) I was appointed as Greek representative to the NATO Science Committee. This is the highest body of NATO on scientific affairs, and it has a substantial budget. Its main subsidiary bodies are: (a) the research grants panel, that supports collaborations of scientists of different NATO countries, and (b) the summer schools panel that organizes several international summer schools in many NATO countries. There are also various smaller panels devoted to subjects of special interest, like biology etc.

Some representatives of various countries were among the top scientists of each country. E.g. USA was represented by the nobelist I. Rabi, France by the nobelist L. Néel, Germany by a minister of science, and so on. I became friend with all the representatives and I succeeded to promote several high ranking Greek scientists as members of various panels. When I started there was only one Greek member in one panel. But during my term, seven top Greek scientists were appointed.

I spent much time with the NATO Science Committee, mainly in travels to Brussels and all over Europe. But in Greece NATO was not liked, especially among the students. There were many demonstrations against NATO in Athens. One could see graffiti in the walls of the University of the form "Down with CoNATOpoulos".

However my participation in the Science Committee was appreciated by the members of the Committee. Thus, when the term of the Chairman of the Committee was over, several members of the Committee wanted me to be the next Chairman. The Dutch representative came to Greece to convince me to take over the Chairmanship. But there was also a proposal for another Greek,

---

[11] Such a judge lived next to my apartment in the same apartment house. One day I noticed a policeman in the entrance, who asked me: "What is your name"? I said "Contopoulos. Why do you ask"? The policeman bowed deeply and said: "At your service, sir". He opened the door for me and escorted me to the elevator. I was surprised. But later I realized that the policeman mistook me for a judge, named Costopoulos, whom he was supposed to protect.

Prof. M. Angelopoulos, who had been a Greek national representative to NATO before me. Dr. Angelopoulos was a good friend of mine, and I supported his candidacy, both in Greece and in the NATO circles. Thus, my answer to the Dutch NATO delegate was that we should all support Dr. Angelopoulos.

But one day the President of the Greek Republic, Mr. Tsatsos, called me and told me that Dr. Angelopoulos was needed for another position, and suggested that I should accept the NATO position myself. This was extremely hard for me. I had four small children and I should move soon with all my family to Brussels. The work of the Chairman required my continuous presence there and, of course, I should stop all my scientific activities. Thus, I said that I could not accept, for personal reasons, and suggested again the candidacy of Dr. Angelopoulos.

If I had accepted, I should be involved in politics. In fact, the Chairman of the Science Committee, is a Vice-Secretary General of NATO, only one step below the Secretary General (at that time Mr. Luns). It was an important position but not for me.

Later on the political situation in Greece changed. After the elections of 1981, the socialists under Andreas Papandreou came to power. Papandreou as leader of the opposition was critical of NATO and it was not certain that Greece should remain in NATO. (Later, he changed his policy and became a strong supporter of NATO). At any rate I felt that my position as representative of Greece was uncertain. Thus, I went to see the new vice-minister of defence, who was responsible for NATO affairs, Mr. Petsos, and gave him my resignation. I told him that I wanted to enable him to choose another person, trusted by the new government, to represent Greece in NATO. At the same time I gave him a report of my activities in NATO.

When the vice-minister read my report he called me and said that he wanted me, by all means, to continue as Greece's representative and promised me full support.

But then I realized that my job was no more easy. I had to meet a number of ministers and discuss matters of NATO with them. All the ministers were very polite to me, but it was clear that they, themselves, did not know what was the new policy of the government as regards NATO. My argument was that as long as Greece belonged to NATO we should profit from the financial and scientific support of NATO, even if later Greece might withdraw completely. But the reaction of the ministers was evasive. It was clear that they wanted first definite instructions from the prime minister before doing anything.

This uncertainty made my position ambiguous. There was a big program in NATO at that time, to support the infrastructure of its weaker countries, Greece, Turkey and Portugal. But Greece could not absorb its share of money. Thus, at a meeting in 1983 the US representative proposed to give this money to Turkey. His proposal was seconded by the United Kingdom and Germany and it was clear that it would be accepted by the Committee. I got up and protested strongly. I said that the delay was due to the change of government in Greece, and they should give us some more time to adjust. Then the Chairman (a French) sided with me and proposed to postpone any decision until the next meeting, six months later. This proposal was reluctantly accepted by the Committee.

When I came back to Greece I tried to meet some ministers to explain the urgency of the situation. But in vain. I realized, later, that the vice-minister for defense himself was in disfavour, and he was later thrown out of the Government. (Later on, this same vice-minister was a target of the terrorist organization 17 November, and he barely escaped death).

Then, exasperated, I submitted my resignation and stepped out. I am sorry to say that later the role of Greece in the NATO Science Committee was considerably reduced.

Several years later, in 1992, I was proposed to NATO as a member of the research grants panel. This was approved by the NATO Science Committee and I started again a collaboration, at a different level. Our main job was to judge and approve collaborative research programs. Such programs involved mainly two different NATO countries. But at that time there was the dissolution of the USSR and of the Warsaw treaty. Thus, NATO decided to accept also participants from eastern countries in joint projects.

We were instructed that we should support all worthy projects, independently of the country of the participants (at any rate one person at least should be from a member country).

We accepted several valuable projects involving Russians. But then the NATO authorities had second thoughts. They said that we should reduce considerably the Russian participation in our proposals. Several people of the panel, including me, protested strongly. We said that we had done our job of evaluation and if the decision now was political, rather than scientific, its implementation should be left to the NATO authorities.

At the same time there were many discussions about a possible NATO action against the Serbs in Bosnia. NATO planes were patrolling this area and there was a possibility of a NATO attack there. I voiced my complete opposition to such a plan. I said that NATO was a defense organization and had no business to intervene outside its member countries. My colleagues were sympathetic to my views. One colleague said: "If one NATO plane, patrolling over Bosnia, is shot down, its cost is several budgets of our Science Committee. NATO can do a much better work to preserve peace by supporting scientific programs, than by dropping bombs".

But the hawks in NATO had the upper hand. The intervention and the bombing of Serbs in Bosnia started in 1994. It was, supposed to be justified because the Serbs had, presumably, exploded a bomb in the central market of Serajevo, that killed many people. This bombing was proved later to have come from the muslim side (west) and not from the serbian positions in the east. But the truth came too late. Meanwhile the bombings started without any mercy, and resulted in the defeat of the Bosnian Serbs, that had nothing comparable to fight with.

When the bombings started I felt that I could not support any more, even indirectly, the offensive actions of NATO. I did support NATO as a defensive organization, and I suffered serious harassments in my University because of that. But if NATO wanted to extend its role outside its limits and outside its original constitution, I could not participate in such actions. Thus, I sent my resignation to the Chairman of the Science Committee on 11 April 1994, as a protest against these aggressive acts. In particular I accused the General Secretary of NATO for an unacceptable propaganda in favour of NATO's actions.

The Chairman of the Science Committee, Dr. J. M. Cadiou (a French), replied to me as follows: "I cannot let pass your unwarranted assertion that NATO has, by its activities in support of the United Nations in Bosnia, changed from being an essentially defensive organization and became an "aggressor". There is no basis for this assertion, or of the assertion that NATO has "taken sides" in Bosnia. All of NATO's actions have been in support of the United Nations and indeed have been in response to requests by the United Nations. NATO as such does not have an autonomous role in the former Yugoslavia. I also entirely reject your unjustified accusations against the Secretary General in particular".

My reply to him was as follows: "I was surprised by your letter of 3 May 1994, replying to my letter of resignation from NATO's research grants panel. You did not answer any of the main points of my letter, demonstrating NATO's aggressive policy in Bosnia, but you only claim that NATO is innocent because it acted "in support of the United Nations". I am sorry for you if such an argument can set at ease your conscience. A similar argument was used by Pontius Pilate 2000 years ago".

It was really naive to believe that the United Nations had the initiative of the actions against the Serbs. This was proved clearly later when NATO attacked Serbia itself, without any sanctioning from the United Nations.

This second phase of attack started with Kosovo. Kosovo was a predominantly Serb province until the second World War. During the German occupation the nazis formed an Ustashi (Croatian) government that exterminated an estimated 1.000.000 of Serbs. Furthermore many Serbs were forced out of Kosovo. When Tito (a Croat communist) took over in Yugoslavia he did not allow any Serbs to return to Kosovo. Kosovo had now a majority of muslims (Albanians).

When Yugoslavia was dissolved many subordinates of Tito took over as leaders of the various countries that resulted. Milocevic in Serbia, Tutzman in

Croatia, Izbekovic in Bosnia, Gligorov in Scopia. All of them claimed not to be communists any more, but nationalists, or socialists.

I had many Serb friends that opposed Milocevic. They considered him a traitor, who had wronged Serbia considerably. In particular he had signed the Dayton agreements that gave away Kraina, Slavonia, and large parts of Bosnia. At that time Milocevic was favoured by the US government, and no serious accusations were directed against him. But when later he did not agree to give away also Kosovo, the whole propaganda of the US government was turned against him as a war criminal.

I did not like Milocevic and I was happy when later, Kostunitsa won the elections and deposed him. But I believed that NATO's agression was not directed against Milocevic, but against Serbia, and that was unjust and unjustified. Thus, I asked several colleagues, presidents of many nongovernmental organizations in Greece to send a protest against the bombings in Yugoslavia.

We sent an e-mail to several hundreds of scientists, or political persons all over the world. We explained some basic facts about the Kosovo-Yugoslavia problem, that we knew, first hand, from friends that had worked or visited Kosovo many times, and we expressed the reasons why we believed that the bombings were wrong and affected the wrong people, and not those that were really to blame.

We received about 500 responses to our e-mail letters. Half of them favourable and half negative. Several people wrote us that they were ashamed by the actions of their governments (U.S., U.K., France, etc).

But others were strongly critical. One even wanted to send us to the international court "for supporting Milocevic". Another "colleague" from a respected international institute simply wrote : "The Serbs are killers. They must all be exterminated from this planet "(sic). Still others, instead of a reply, wrote us the worst possible abuses.

The end of the story is well known. After the occupation of Kosovo by the NATO troops they expected to find over one hundred thousand graves of killed Albanians. But, despite the enormous propaganda, they could not find more that a few hundred killed, and it was not clear which of them were Albanians and which Serbs. After the "liberation" of Kosovo some thousands of Serbs were killed as a retaliation by the Albanians, more than the victims of the previous period.

During the war in Kosovo, several people claimed that our attitude was biased because the Serbs are orthodox christians like the Greeks, while the Albanians are mostly moslems. But even stronger opposition in our country was recently directed against the war in Iraq, which is a completely moslem country.

My association with NATO ended completely a few years after the war in Kosovo. In 1999 I had been invited to a NATO summer school on dynamical astronomy. I had accepted, and even sent a summary of my invited talk. (I was invited to all NATO schools on dynamical astronomy since 1970).

But when the school was finally organized my name did not appear among the invited speakers. Even more, all Greek invited participants were deleted.

I had the opportunity to ask personally the organizer, why this happened. He tried to give an excuse, that they had too many invited speakers, etc. But when I asked bluntly: "Did you get instructions from NATO to take out Contopoulos and all Greek invited speakers?", he only protested weakly; but then he burst out: "Politics is very bad when it interferes with science" he said.

It is a pity that the NATO meetings, that even during the cold war years were encouraging people from the eastern countries to participate, had now become so politicized, that unless someone followed the official policies of NATO he could be put away, despite his scientific status.

I was surprised by the attitude of several of our colleagues abroad. During the Kosovo crisis, there were people telling us: "We do not believe you, because we heard the CNN saying the opposite". But CNN in this and in many other crises has been extremely one-sided. Even if they reproduced broadcasts from Belgrade they repeatedly warned people that these represented the biased views of Milocevic, while they did not give a similar warning when they broadcasted the views of NATO. Thus, the bombing of a train that was going to Greece (!), the bombing of the Chinese embassy in Belgrade, etc, were simply "mistakes". The killing of 75 Albanian refugees in a convoy in Kosovo was first attributed to Serbian aircraft (nonexistent at that time). Later it was realized not only that a specific NATO plane had done the bombing, but that it was ordered (in real time) to do so. We heard the taped discussion of the pilot with his superiors during the raid, who ordered him to attack the convoy. Later it was accepted that all these raids were done on purpose. "But, they said, some lies are necessary during a war".

On the contrary the British BBC had a much more open attitude. It presented both sides, and although it condemned Milocevic, it allowed us to see also the destruction of innocent people, both in Kosovo and in Serbia.

I do not expect my colleagues to follow one policy rather than another, but I do expect them to require to know the facts. They should not believe any propaganda at face value.

The effects of propaganda are not to be found only in totalitarian regimes, like the nazi Germany, and the Stalinist USSR. One may remember also the MacCarthy period in the United States, the support of the dictatorships in South America, etc.

It is unbelievable to what degree the U.S. media influence public opinion with the most simplistic slogans. The world is separated into bad guys (our enemies) and good guys (us, of course) and there is no intermediate position.

The only hope in that country is the reaction of many intellectuals, who can have an independent opinion. But even among the highest levels of intellectuals one finds the underlying simplistic attitude (good guys against bad guys). I remember what happened in Washington in 1975 when Nixon resigned after the Watergate scandal. I was at a reception in the Australian embassy, in downtown Washington, when someone told me: "Let us go home. Today Nixon is probably going to announce his resignation". After Nixon's resignation all my friends were celebrating. I was at a party with several colleagues. They were congratulating themselves that this "bad guy" was forced to quit. Among the general exuberance I made a sobering remark. "It is not only Nixon who says lies. Remember the U.S. promises to many people abroad (and I mentioned Cyprus and Palestine), that were never fulfilled". There was an awkward silence among my colleagues and then someone said: "I know that we lied to many people abroad. But we do not care. We only care about lies to the American people".

Fortunately such a cynical attitude is rather exceptional among our American colleagues. Most of them are idealistic, as I know from personal experience. I will mention only one case. I was at the closing dinner after an international meeting and I was sitting in a table with several American participants. After some polite remarks I asked them how they felt after the Bush administration's attack to Iraq. All of them felt indignant, especially because of the disclosure that there were no weapons of mass destruction in Iraq. But then one lady said: "Please speak lower. Someone may hear us". I was surprised. "Are you afraid"? I said. "Why"? And the lady explained. "There are many people that have been arrested, as being unpatriotic. So we must be careful". Then I burst out: "Is that the kind of democracy that your government wants to impose upon the world"?

52. MOTIVATION

The last sections of this autobiography are devoted mainly to some general topics. The first refers to the question: "Why people study astronomy?" When I was finishing high school and I was considering what career I should follow, I wanted to study the whole universe. From elementary particles to cosmology. I realized (correctly, as it turned out) that in order to do that I should first study mathematics. That is why I went to the school of mathematics of the University of Athens. Although I had succeeded also in the entrance examinations of the engineering school (a most prestigious school at that time) I left it for the school of mathematics. When I went to take over my papers from the school of engineering, the secretary could not find my file. "I am sorry, he said, but I looked at all the files of those that failed and I cannot find your name". "But I did not fail, I said, I was successful". He looked very astonished. He found my file, saw my identification papers, but still he could not believe that I was serious. Finally he said: "Please write me an affidative that you yourself, of your own free will, take away these documents". And only when I did that he gave me my papers.

Many years later I realized that I could not do everything in science. I had to concentrate on some subjects and just read about the rest. Nevertheless I was happy to find a broad enough subject, namely dynamical astronomy, that spanned physics and astronomy from molecular physics to cosmology.

That was a good portion of my original ambition. But of course I had to concentrate on particular topics in order to find really new results.

My career was rather peculiar, because I did not follow formal graduate studies. Most of my knowledge of mathematics, physics and astronomy was due to my reading of many books without any guidance. That was unfortunate because a correct guidance would save me much time. But on the other hand it allowed me to explore subjects that were not fashionable at that time (in the 50's), but proved later to be very important.

For example I read a lot of stellar dynamics and celestial mechanics at the same time. These two subjects seemed very remote from each other. People in celestial mechanics were calculating orbits but did not consider the statistics of orbits and did not use any methods of statistical mechanics. Similarly, people in stellar dynamics used statistical mechanics profusely, but did not care about the details of stellar orbits. But I found that a combination of the methods of celestial mechanics and stellar dynamics could open new possibilities of research. Later the subject of orbits and integrals of motion in dynamical systems, proved to be of great theoretical interest and with many practical applications.

At the same time (in the 50's) a new important development was the emergence of computers. My interest in computers started around 1955 and my first calculations were made in 1956. The use of computers in dynamical astronomy proved to be a source of unlimited possibilities, allowing the rapid expansion both of the theory of dynamical systems and of its numerous applications.

I followed the development of the new techniques of computers as soon as they appeared. First we used typewriters that typed the results in real time (but incredibly slowly). Later we used punched cards, tapes, and finally windows. The speed and capacity of computers increased enormously over the years. Various accessories, graphics, colours, etc. were also improving continuously.

During my early years with computers (in the late 50's and in the 60's) I had to overcome several problems. For example I found a persistent error in some of my calculations that was due to an error in the Fortran package (the Fortran had not foreseen what would happen if zero is raised to a power zero). An assistant of mine was doing extensive calculations with long decks of punched cards and a damaged card produced an error that invalidated all his work, that had to be repeated.

Nevertheless I tried every innovation in the computers, e.g. in using graphics. In the early 60's I even produced one of the first computer movies in astronomy. The director of the Institute for Space Studies in New York, Dr. R. Jastrow, encouraged me to do it, and provided the technical help for that. The movie was extremely simple. It only represented a moving point describing an orbit.

But in the last years I gave up trying to follow the latest innovations. My assistants now are much better informed about the possibilities of modern computers. They do simulations of N-body systems with N of the order of one million, on a desk computer, while in the past only with the use of Crays we could do such calculations.

My experience after 50 years of research, allows me to give some advice to my students that may be useful. In order to be successful in scientific research four things are necessary:

1) Enthusiasm for research,

2) Not to be afraid of much work and innovation,

3) An appreciation of one's limits,

4) Honesty, as a pre-requisite of all scientific work.

Let me develop these four items a little further.

1) Many students come and ask to work with me. But in several cases their interest is only to have a good career, or to get a particular job. Although I want to help all my students, when I disclose such an attitude, I tell them that it would be difficult for them to do good scientific research. I remember one such student. I told him that he might do better in computers than in astronomy. And I gave him some addresses abroad and several recommendation letters. Some years later he came to thank me for my advice. He was already rich, with a salary in a computer company higher than my own salary as a professor, and he was quite happy.

On the other hand life in science is often difficult. Normally one never expects a leisurely life. I never had such a leisurely life myself, although I had quite high positions in academia and I worked extremely hard for 50 years.

Furthermore, one often has disappointments in his research, either failures in his efforts on a particular problem, or in getting a grant for his further work, or a position, that he believes that he justly deserves. One needs an important underlying reservoir of enthusiasm to be able to overcome these difficulties.

2) The work required for research is really enormous. First, one needs to absorb a large number of books as a background in mathematics, physics, astronomy, etc. If one finds a good advisor, then he may go directly to the best books on each subject. But he will always find new books to read throughout his life.

Then one will need to read a very large number of papers. Not only papers closely related to his present work, but also papers on broader and/or related subjects. The number of papers is so large, that it is never possible to read carefully all of them. But one has to understand the basic points of a paper and its close or remote relation to his main subjects of interest.

The reading of papers is in general quite difficult. Part of the difficulty is due to the authors, especially if they fail to give the proper definitions and only use their esoteric specialized jargon (see section 53). But even if the authors are clear, they usually look at various problems from a different perspective from ours. Thus, one has usually to learn quite a number of new notions, new methods, new mathematics, new physics, in order to understand these new aspects of a problem. But such a study is really rewarding. One learns many useful new things concerning his own problems that may help to generalize and extend his results in an unforseen way.

For example the problems of dynamical astronomy can be considered with various mathematical tools, like ordinary and partial differential equations, topological methods, ergodic theory, combinations, integral equations, numerical methods, etc. Such problems may have many applications in physics and astronomy, like classical mechanics, continuum mechanics, quantum mechanics, chaos theory, celestial mechanics, galactic dynamics, cosmology, etc.

In order to be able to follow the various aspects of a problem one should not be afraid of innovation. Many people stop in front of a paper with an expression like "too mathematical", or "too abstract", or "too numerical" etc. But if one insists one finds sometimes (not always) some gems that are really worth noticing.

I cannot give an algorithm what papers to read and what to leave aside. But I found many interesting papers in journals far away from my own circle of interests. That allowed me at least to give a rather extensive list of references in many of my papers and prepare the diverse contents of my book on "Order and Chaos in Dynamical Astronomy" (sections 35 and 55).

Sometimes people try to work in areas for which they have not the right preparation. Then they try to use the methods they know in a new field without due consideration of other aspects with which they are not familiar.

I remember one well known physicist who was trying to find the physical basis of consciousness. He was giving a lecture on the dynamics of the brain, using quantum mechanics. In the question period a colleague asked him why he considered quantum mechanical phenomena, while the classical theory of chaos might give a better explanation. His answer was really astonishing. He said that one applies the physics that he knows best, and in his case that was quantum mechanics (as if the brain was forced to use quantum mechanics just to satisfy him).

Finally, one has to do a lot of work on any particular subject one studies. In the case of analytic work one has to make a lot of efforts until the final result is reached. The final text may be short, but the preliminary analytical work may be quite extensive. In the case of numerical work one usually does many times more calculations than finally published. This work is necessary in order to understand the underlying structure of the problem. But in the printed version it is sufficient to include some representative results. In other cases one may do several numerical experiments and summarize the findings with only a few words.

This "wasting" of efforts is necessary in order to have a real progress in science. When an explorer discovers a new continent, he should not try to give immediately all its details but only a rough map, to guide further navigators that sail in these waters.

3) Most people working in science are very ambitious. But unless one knows his limitations he is destined to fail. One cannot discover the whole universe with one strike. Of course he should use his abilities to the maximum. But if he attacks a problem much more difficult than what he is reasonably expected to solve, then he will be disappointed.

In some cases this disappointment is due to a supervisor, who asks a student to do what he himself has been unable to achieve. That is not only foolish, but also ethically blameful. Only if one is sure that a solution is close at hand he may ask his students to try to find it.

But in other cases the blame is to be put on the student. I had a very good student, who came to ask me where he should do his PhD. He had got scholarships from MIT and Cornell and he should make his choice. I asked him what he wanted to do. He said: "The theory of everything". I told him that this was not a reasonable subject for a PhD. Such a theory might come out after a whole life of research. But may be a theory of everything does not even exist. I suggested to him to do a more reasonable PhD and work extra, in his spare time, on a possible theory of everything. But my student insisted. He said that nothing inferior should satisfy him, and he should either succeed or perish. I am sorry to say that this student failed in his career. Many years later he had not finished his PhD and finally I lost contact with him.

Another Greek student was working on a similar subject, namely the theory of superstrings. He worked under one of the main proponents of the superstring theory. But although he was very bright he could not go much further than his supervisor. However this student was clever enough to do a drastic change. He left his supervisor, went to another university, and did his thesis on a very different topic.

Thus, one should be able to see not only his own limitations, but also the limitations of his supervisor. Of course it is not an easy decision for one to change his supervisor. But if necessary this also must be done.

4) Perhaps I should not mention honesty as a prerequisite of scientific research, because this is a prerequisite of all human activities. But in science honesty is so important that it should be (and it is) stressed to the extreme. I tell my students that science could not exist without honesty, without the certainty

that a recorded fact is a fact. Even mistakes are hardly tolerated, and only if they are not numerous.

Therefore, a break of honesty is punished by the capital punishment: (scientific) death. No one takes notice of the "results" of a person who has been caught in a fraudulent behaviour.

I had such an unfortunate case in my department. One of my assistants was assigned to participate in the solar flare patrol. He filled daily reports with his observations that lasted a few hours every day, and he sent them to an international center in the U.S.A.

For some time this work went on without any problems. But one day I received a letter from the international agency. "You report a flare on that day and time. But others observing also the sun at that time did not see a flare. Why?" I asked the assistant what happened. He mumbled that perhaps he had mistaken a brightening for a flare. But a few days later I received similar letters with the same question. "You mark a flare where no flare appeared". This was not a statistical mistake. I insisted that the assistant should tell me the truth. I told him: "If you do these observations and you mistake a small brightening for a flare then you are a bad observer". Then he reluctantly accepted that he had asked the doorman (a quite inexperienced person) to stay on the telescope and report to him when he saw flares. I immediately announced that our flare patrol was stopped, and did not allow this assistant to send any further reports.

The Greek students that receive their PhD give a pledge similar to the Hippokrates' oath of the medical doctors. It is written in a dignified style in ancient Greek. Among other things the prospective doctor promises "not to teach anything that is opposite to my own knowledge, and not put to shame my own dignity as a disciple of the Muses by a disorderly conduct". This pledge could well be a leading principle for scientists all over the world.

53. PAPERS AND REFEREES

When one finds some new interesting results, he must present them in an appropriate way.

I am not happy with the way I wrote some of my early papers. Their drawback was that they mixed the essential points with a lot of mathematical details, which should be placed in Appendices. As an example I mention my paper "Resonance cases in a third integral" (Contopoulos 1963b). In this paper I introduce a general method to find integrals of motion in resonance cases, something novel and important, in view of its extensive applications. But the basic idea is hidden behind a lot of algebra. Despite its drawbacks it is cited by about 80 people, but it could have a much better impact if the main results were more clearly emphasized.

Another error I made when I published a paper containing "Tables of the third integral" (Contopoulos 1966b). I was impressed by the ability of the computer to prepare these Tables directly for publication. But such Tables are never used. I should have published only the text of this paper, which contains valuable information, and simply note that the computer program is available on request.

Gradually I learned a more "correct" way to write papers. I realized that the reader looks first only at the title and the abstract of the paper. If this is interesting enough he looks at the conclusions and the figures. Only if he is particularly interested in the subject he tries to understand the main points of the paper. Then a senior scientist usually gives the paper to a younger colleague to read it carefully.

It is essential to emphasize the main results of a paper. But one should also avoid the other extreme, i.e. to mention only the results, without sufficient proof, or justification.

Many authors make claims that are not sufficiently justified. In particular the mathematical derivations in many cases are of the form "After some (or many) operations we find....". If one has enough experience with calculations of this type he may guess what operations have to be made, otherwise one has to accept the result by faith alone.

I always stress that there must be at least one person in the world besides the author that should be able to understand all the details of a given paper. This person is in principle the referee. The referee should be able to repeat the calculations of a paper if he wants. When I referee a difficult paper I sometimes ask the author to provide detailed mathematical proofs, even if he may not want them to be published.

In Chandrasekhar's famous book "The Mathematical Theory of Black Holes" (1983), it is stated that the details of certain long calculations (600 pages!) are deposited in a particular library of the University of Chicago for interested readers.

Of course the referee should not be asked to repeat the numerical calculations of a paper. But an experienced referee should be able to suspect possible errors and indicate them to the authors.

One of the best referees that I have met was Michel Hénon. I remember that in one case he found an error in the numerical results of a paper by one of my assistants. He pointed out that a particular characteristic (a curve giving the position of periodic orbits as a function of the energy) consisted of two parts. My assistant was astonished. "How could he have found such a small gap between the two curves?" (In fact my assistant had calculated only a few points along the curve and joined them by a continuous line without a closer search).

I have refereed several hundreds of papers for the main astronomical journals, but also for many journals in physics, like the Physical Review, and the Physical Review Letters. I am not an "easy" referee, but I always try to make constructive remarks, especially in cases of papers from isolated countries, like the Soviet Union, or China, lacking good libraries, and authors having difficulties with the language.

There are over 300 people thanking me in their acknowledgements for improving their papers (and I do not mention those that changed their papers considerably without thanking me). Among them there are some of the best papers in galactic dynamics and related fields. An example is the following reply of a particular author: "I have been publishing papers for three decades now and I cannot remember receiving such expert and author friendly refereeing".

I could mention many papers that site my contribution in the acknowledgements. A characteristic remark was made by P. Magnenat, who did his thesis in Geneva: "My thanks go also to Professor Contopoulos for the precious suggestions that he has given me whenever he was passing through Geneva". In fact I could help in a similar way several young scientists during my visits to various institutes abroad.

In a particular case I received a paper from the Soviet Union that had some good ideas but was badly written. I made a number of suggestions about the scientific content of the paper, and I added that its language should be improved considerably. When the paper came back to me its contents were more or less correct, but its English was still bad. I replied that the paper should be accepted only if the language would be improved. Then the authors wrote to the editor: "Sir, we have not studied in Oxford, or Cambridge. Therefore the remarks of the referee about the language of our paper are unethical (sic)"! I wrote to the editor that if my remarks were considered unethical, he should send the paper to another referee. The second referee wrote: "I cannot understand this paper at all. Reject it!". Then the authors wrote to me: "Would you be kind enough to correct, yourself, the language of our paper?" I had to rewrite the whole paper (people told me that I should be a co-author) and after that the paper took an acceptable form. The authors thanked me with many warm expressions in the acknowledgements (Gurzadyan and Savvidy 1986).

Unfortunately the refereeing system is very inhomogeneous. I am sorry to say that several referees look mainly at the list of references to find their name. The paper is accepted after some of their own papers are added in the list. There are also cases of conflict of interests. In one case I had to referee a paper on the third integral in a spiral galaxy. I was finishing a paper on the same subject myself at that time. At first I thought of refusing to referee this paper. But the paper was good (I had only a few remarks). Thus, I suggested to be accepted for publication. My own paper, that was a sequence of my earlier papers on the third integral, appeared a few months later (Contopoulos 1975) and I took care to avoid much overlapping.

But not all referees act in the same way. In one case a paper of mine was delayed by the referee for two years with various excuses, although his remarks were of quite minor nature. After two years the referee published a paper of his own on the same subject, and the overlapping with my paper was serious. Thus, I decided to withdraw my paper. It is ironic that after that the referee wrote me a personal letter saying: "Why don't you publish your very nice paper?"

I remember one case where there were two referees, M. Hénon and myself, for a paper by J.H. Bartlett. The paper was supposed to contain numerical evidence that there are no invariant curves around a stable invariant point. It calculated empirically some curves that looked like invariant curves (tori), but they were only cantori in modern nomenclature (section 46). Orbits from the interior of such a curve could go outside it after a long time. Therefore the empirical curves were not real invariant curves. I wrote that this evidence was not sufficient to exclude all invariant curves. In fact according to the KAM theorem close to the invariant point there are real invariant curves. Therefore, the numerical exploration should try to distinguish between real invariant curves (tori) and pseudo-invariant curves (cantori).

It is of interest that M. Hénon had written, independently, very similar remarks to the Editor. But his conclusion was the opposite to mine. I wrote: "The paper is wrong. Reject it!", while Hénon wrote: "I believe that the paper is wrong. But let the astronomical community decide about its correctness or non- correctness". Thus he suggested publication of this paper!

I had later several personal discussions with the author. Although his conclusions were wrong, Dr. Bartlett had done serious numerical work. He had done many detailed and careful calculations, including the first numerical calculations of heteroclinic orbits. Thus, I invited him to one of our nonlinear workshops in Florida, and I am happy that he published there a paper where he gives essentially the limits of the real invariant curves in a particular system (Bartlett 1987).

In fact Bartlett found evidence both for real invariant curves (KAM curves) and pseudo-invariant curves further away (Bartlett 1989 and later work). In this paper Bartlett characteristically thanks "G. Contopoulos for a stimulating discussion on October 7, 1987 (sic) which provided the impetus for the ... present results" (Bartlett 1989).

But such friendly disagreements are not very common. Sometimes the criticisms of the referee are so violent that they express his own bias. In one case I submitted a paper discussing a particular dynamical model and I simply stated that my model did not belong to the class of models considered by Dr. X. The referee discussed at length the importance of the models of Dr. X. (as if I was the referee of his paper) and concluded his report by saying: "I respectfully request (sic) that this paper (by Contopoulos) should not be accepted". But a second referee was quite positive and my paper was finally published.

It is reasonable, if a paper disagrees with the conclusions of another author, that the editor should ask the opinion of this particular author. But the final decision must be based on the opinion of another independent referee.

In one particular case a paper attacked a previous paper of mine. I wrote a protest stating clearly that the first author had misunderstood me, and I sent it to the same journal. The editor sent it to the first author, who insisted in his view. I replied again. But the first author wanted to publish another paper as a reply to my protest. Then I asked the editor to send all the correspondence to an independent referee. The result was that my protest was published, without any reply by the first author.

When I feel that I am not competent in a particular field, I do not accept to referee a paper in this field. But many people do the refereeing although they are not competent. In several cases it is obvious, from the statements of the referee, that he is not familiar with the subject. Then why does he accept to referee such a paper? An incompetent referee is harmful both when he rejects and when he accepts a paper. I think that the refereeing system should be improved by eliminating the anonymity of the referees. The referee should feel that he is doing a responsible work under his own name. He should know that his judgement will be judged by others. Even the most conscientious referee will be more careful in his remarks (negative or positive) if he knows that his remarks will be known.

The choice of appropriate referees is the most difficult task of the editor. Referees that are extremely slow, or obviously biased, should be excluded. If there are two competing groups in a certain field the editor often sends a paper from one group to be refereed by the other group. But then the editor must be careful, to avoid unfair judgements. He cannot accept all criticisms without a sufficient justification. For this reason the editor is "the referee of the referees".

Nevertheless, despite its limitations, the refereeing system is useful. It is essential in keeping high standards in our scientific journals, if the editors try to impose the appropriate high scientific and ethical standards. The scientific community must show its appreciation for the work of the editors and of the referees. Judging from myself, I find that it is quite a burden to make good refereeing for more than 20 papers per year as I do.

I remember that I had some informal discussions with our editors of the journal Astronomy and Astrophysics about papers from countries like the Soviet Union and China. I suggested that we should help people from these countries to publish in our journal. One editor replied: "Help, yes. But not at the cost of the quality of the papers. Otherwise these people will never reach the international level required". He was quite right of course. Thus, our help was indirect, e.g. by waiving page charges, providing some translations, and patient referees, to improve the language.

Outside the refereeing system one receives a lot of useless, or wrong publications. Very often I receive articles, or drafts of articles, or even books, about some fashionable subjects like cosmology, unified theories, philosophy, space science, environment, astrology, etc, not to mention the perpetuum mobile, or the squaring of the circle and other mathematical curiosities. I do not have the time to find what is wrong with each publication of this kind, but even if I find it, the interested person is never convinced.

On one occasion I received the draft of a large book covering almost everything, cosmology, physics, chemistry, life sciences and philosophy. The author was a professor in a small American college. His book was sent by various publishers to a number of prominent referees, including Linus Pauling, but all of them either rejected the book, or wrote that they did not feel competent to judge it. The author sent me his manuscript through a friend of mine and asked for my support. I tried to make some useful suggestions about the cosmological part of the book, that dealt mainly with various numerical relations involving the large numbers of Dirac. My main suggestion to the author was to publish first a number of papers on particular topics, to see the reaction of the related scientists, and then assemble everything as a book. But I wrote also that I disagreed with some of his conclusions. His reaction was vehement. He attacked me of being tied up to the scientific establishment, and not allowing the progress of science outside the system. I am only sorry for the time I spent reading this book.

But coming back to the real scientific literature, I find often a peculiar drawback in many papers. The lack of an explanation of the symbols and of the terminology used. The authors are often so used in their jargon that they do not realize that others do not have the same background. I consider the lack of definitions as a serious drawback that makes several papers unnecessarily difficult to read.

Another drawback of many papers is the lack of the appropriate references. Some people do not care to make a search for the relevant literature. This is particularly true for many American authors that do not even look at other journals besides the Astrophysical Journal (section 25). But not only Americans.

In one case I refereed a paper on "Spectral stellar dynamics" (Binney and Spergel 1982). I made the remark that similar methods were already developed in the chemical literature, and I gave a particular reference. Dr. Binney replied that when he found his results he had the impression that he was rediscovering the wheel. However, when he spoke about that to his colleagues in Oxford and in Princeton no one knew anything about that. Then the published paper has the following acknowledgement: "We thank the referee, G. Contopoulos, for helpful comments on the manuscript and for drawing our attention to papers concerned with spectral dynamics in the chemical literature".

The fact that spectral methods were already used in chemistry did not reduce the importance of their applications in galactic dynamics. This subject became even more fashionable after its extension in the form of "frequency analysis" by Laskar and his associates.

A very curious mixture of attitudes as regards referencing was to be seen in the Soviet Union. Some people referred only, or mainly, to Soviet papers (perhaps they had difficulties with foreign languages). But others made very careful references to papers all over the world (American, European, Japanese, etc.). I remember a paper that compared my work to that of M. Kruskal, a leading professor of mathematics in Princeton (Ossipkov 1977). When I met Kruskal I gave him a copy of this paper and I told him: "It was a special honor for me to be associated with your name". And Kruskal very politely replied: "The honor was mine".

As a conclusion, I would like to suggest the following to prospective authors:

1) Write in English. I do not think that English is the best language for science, but English is established. Personally I would prefer French or German. However French and German are not understandable by most readers.

2) Separate clearly what is essential in a paper from the detailed proofs and discussions. Whatever is secondary in the proofs and in the discussions may be put in Appendices.

3) Give clear definitions of all the terms and symbols used.

4) Give good figures. These may provide the essential points of a paper, in a most clear way.

5) Give the correct references. Many people think that they have discovered the wheel, but the wheel exists since several thousands of years. In the same way there are many re-discoveries of Poincaré's theorems that are already included in his "Méthodes Nouvelles de la Méchanique Céleste". But do not worry. There are many more things to be discovered today than in the past.

54. LECTURES

Similar considerations apply to the lectures, seminars, and colloquia. I gave many lectures all over the world, mainly in the United States, but also in Canada, Chile, Argentina, England, France, Spain, Portugal, Belgium, Netherlands, Germany, Denmark, Sweden, Finland, Switzerland, Austria, Italy, Yugoslavia, Slovenia, Hungary, Czech Republic, Slovakia, Poland, Russia, Georgia, Turkey, Israel, Australia, and, of course Greece.

The main problem in giving lectures is to know the level of the audience. I gave lectures to specialists in galactic dynamics, or in chaos theory, but also to more general audiences of astronomers, of physicists, or of mathematicians, review lectures for scientists in general, and popularizing lectures for the general public, or even for high school children. But one has to speak at the appropriate level to every audience. I remember that I gave essentially the same lecture to astronomers and then to physicists. The reaction was enthusiastic in the first case and reserved in the second case.

In the course of time I made several mistakes. On one occasion I gave two lectures in Boston, one at Harvard and the other at MIT. The first was addressed to nonspecialist astronomers. I made it as simple as possible and it was quite successful. The second was addressed to people working in dynamics. Thus, I assumed that the audience were familiar with many notions of dynamics. But this was a mistake. Some people started from the very beginning interrupting me: "What do you mean by that and that?" until I realized that I should explain everything in detail in order to be understood.

In principle it is preferable to give a simpler talk, even if there are some specialists in the audience. In one case a lecturer asked the audience: "What should I assume as known in this field?". A senior person from the audience replied: "You may assume that we know nothing. But we learn fast."

Another mistake of my earlier lectures was that they were overburdened with mathematical formulae. I remember one such occasion in Seattle. I entered the big lecture room before the lecture and I saw that there were big blackboards covering all four sides of the room. To save time I filled all the blackboards with formulae, and then I was going around the room explaining these formulae. But this sequence of formulae was boring for most of the audience.

Nowadays the use of viewgraph projectors has simplified the task of the lecturer. But the viewgraphs have to be clear, with large letters, and not much information on each of them. Furthermore, it is important to check if the appropriate facilities (viewgraph projectors, slide projectors and blackboards) exist. In some cases there is no blackboard and people are supposed to write on the viewgraphs themselves. I remember one such case. Someone asked me a question and instead of writing on the viewgraph, I automatically went to the projection screen, and started writing on it, until I heard a loud "No!" from the audience.

On another occasion I was supposed to speak to mathematicians. I entered the lecture room with my viewgraphs and I asked: "Where is the projector?" The president was embarrassed. "We have no such thing" he said. "We only use the blackboard". That case was exceptional. In fact, most lecture rooms have a viewgraph projector available. But some people come with a movie, or a video, and ask for an appropriate movie or video projector the very last moment. That is not always available.

I take this opportunity to relate the style of the lectures of a few well known astronomers.

The fastest speaking person that I met was the Japanese celestial mechanician Y. Hagihara. Hagihara (1970) wrote a modern, but already classical book on celestial mechanics, consisting of five large volumes. This book contains everything connected with celestial mechanics, including the most recent developments in related fields. Among celestial mechanicians it is accepted that the importance of every person in this field is proportional to the number of pages devoted to him in Hagihara's book! (My own place is not bad, despite the fact that most of my work is not in celestial mechanics proper, but in galactic dynamics, and nonlinear dynamics in general; there are 37 pages of Hagihara devoted to my work).

Hagihara was speaking very fast and writing on the blackboard at the same time. He was writing with his right hand, and erasing the blackboard continuously with his left hand to make room for more formulae. But this did not make him easy to understand.

Another peculiar speaker was B. Strömgren. He usually filled the blackboard with numbers and projected images with tables of numbers and we had to try hard to absorb the information provided.

The American speakers are used to say many jokes during their lectures. I remember Leo Kadanoff, who was giving a lecture at Cornell. He was speaking about period doubling bifurcations, when the projector broke down. Kadanoff did not lose his temper. Until the broken lamp was replaced he started saying jokes that made the audience laugh continuously. At the end I asked him a related, but rather difficult question. He stopped for a moment and then said: "Dr. Contopoulos, you know that we have put a lot of things under the rug. Please don't ask such questions". And he did not answer my question.

But jokes cannot fill the lack of scientific content. I remember one speaker that was making continuously jokes. He even made fun of the questions from the audience. At the end most of the audience were angry with him.

Another speaker, from an eastern country, had a huge collection of viewgraphs. He spoke in English (with a strong accent) interspersed with German words. The audience had no time to see the viewgraphs, before they were replaced by others. I told him that he should reduce the number of viewgraphs considerably. At his next lecture he used about 1/3 of his usual number. Then he came to me and asked: "Was that better?" "Yes, I said, it was an improvement. But it would be still better if you reduced the viewgraphs by another factor of two". His third lecture was almost perfect. We had time enough to see what was on the viewgraphs. Then my friend admitted that he did not feel confident in his English, and because of that he wanted to give most of the required information in the form of viewgraphs.

The extreme case of this type was a scientist from an eastern country, which did not speak any English, and only presented a series of viewgraphs without saying a word.

55. ACADEMY OF ATHENS

The Academy of Athens is the highest institution in Greece. It is considered to be a continuation of the Academy of Plato, of ancient Athens, and for this reason it is accepted as the first Academy among all the Academies of the World. Its membership is quite limited. Usually there are less than 50 members from all disciplines (Natural Sciences, Medicine, Arts, Humanities and Social Sciences).

To be a member of the Academy of Athens is the highest distinction in our country. There is only one astronomer, one or two mathematicians, one or two experimental physicists, etc.

I was honored to be elected member of the Academy of Athens in 1997 (February)(Fig. 36). This happened a few months after I became emeritus at the University of Athens (September 1996). At the same time I took over the position of "supervisor" of the Center of Astronomy of the Academy.

This Center is very active. It has now about 20 persons, including 7 graduate students working towards their PhD. Its main interests are: dynamical astronomy, including celestial mechanics, relativity and cosmology, solar physics and solar-terrestrial relations.

We have a weekly seminar that is well attended by specialists from various institutions in Athens, and several people from Greece and abroad have given lectures there.

Our most recent activity (2002) was a workshop on "Chaos in Galaxies", with several participants from 25 countries, all over the world, including people from Japan, Australia, and Chile. Among them were D. Lynden-Bell (England), P.O. Lindblad (Sweden), A. Fridman (Russia), D. Miller, H. Kandrup and C. Hunter (USA), D. Pfenniger (Switzerland), L. Galgani and G. Bertin (Italy), H. Dejonghe (Belgium),P. Grosbøl (ESO), O. Gerhard (Germany), A. Bosma and E. Athanassoula (France), R. Sancisi (Netherlands), B. Jones (Denmark), R. Dvorak (Austria), etc. (Fig. 37)

The scientific program was full of high level lectures. The main new aspect covered by this workshop was the combination of theoretical and observational methods to detect and describe chaos in galaxies. The proceedings were published by Springer Verlag (Contopoulos and Voglis 2003).

The participants enjoyed the hospitality of the Greek sponsors, that was quite substantial.

My involvement in the Center of Astronomy is quite time consuming. At the same time I am teaching a graduate course on "nonlinear systems" at the physics department of the University of Athens. All that is besides the regular meetings of the Academy, the participation in committees, meetings and lectures in Greece and abroad.

Since 1984 we live in a suburb in the Penteli mountain north of Athens. When we first came here there was a very nice forest near our home that covered the whole Penteli mountain. Very often we were walking in this forest. Later, however, a number of forest fires destroyed our forest completely. The worst fire occurred in 1995. I was in my office when my daughter called me to tell me that she saw a fire at the top of the mountain many kilometers away from our

Figure 36. My inaugural lecture at the Academy of Athens.

Figure 37. The participants of the Symposium "Order and Chaos in Galaxies" (2002) in front of the Academy of Athens.

house. I rushed home and I saw the fire progressing very fast. A very strong wind helped the progress of the fire. My wife and I decided to stay at home to fight the fire if it came close to us, and we sent our children away with our car. But when the fire was burning the trees on the left and on the right of our house our children came back with our car and said: "Either you come with us, or we will stay with you". Reluctantly we decided to go away. We were very anxious but fortunately the fire did not harm our house. It burned everything around it and continued further away. After several hours we could return home.

Gradually new trees started sprouting from the ground. But then a second fire in 1998 destroyed what was left. This time the fire developed between Athens and Penteli and all the roads leading to our house were blocked. I tried to proceed by foot but this was very dangerous. So I decided to drive through the burning forest. It was really scary, while the burning needles of the pine trees were falling continuously on my car. Fortunately after a few hundred meters there was a clearing and close to our house there were practically no trees to burn any more.

After this second fire no natural regeneration of the forest could take place. Nevertheless they planted new trees, and our neighbourhood is still nice, as we are in a clear region above the plain of the city.

My most important activity during the last 3 years (2000-2002) was the writing of the book "Order and Chaos in Dynamical Astronomy" that was published by Springer Verlag (September 2002)(Fig. 28b). This book, of 625 pages and over 300 figures, contains the main results of my research work over the last 50 years. Of course it is not so detailed as my original papers, but it contains the main points of my work in many fields (stellar and galactic dynamics, celestial mechanics, relativity and cosmology, and the theory of chaos).

The book deals with order and chaos in general (integrable systems, formal integrals, periodic and non-periodic orbits, systems of 2 and 3 degrees of freedom and N-body systems), and applications of order and chaos to galaxies, to the solar system and to relativity and cosmology. It contains a selected set of 1200 citations and a list of 72 topics for further research. These topics can be used as theses' subjects. In fact, there are now many young people all over the world, working on order and chaos in Astronomy.

I have devoted this book to S. Chandrasekhar. For many years Chandra was suggesting that I should write a book containing my own work in galactic dynamics. He emphasized, correctly, that my work, spread over hundreds of papers, would be lost unless published in the form of a book. He himself, after writing many papers on a particular subject, he would write a book containing his main results on this subject.

Chandra was particularly interested in the field of chaos in his last years. After finishing his monumental work on "The mathematical theory of black holes"

(1983), he spent some years in studying the original work of Isaac Newton (Chandrasekhar 1995). However, at the same time he was reading books on order and chaos and on every possible occasion he was asking me questions about this topic. I am sure that if he lived longer he would have started working seriously on this new topic.

Chandra had expressed his interest on a possible book of mine to Dr. Beiglböck, of Springer Verlag, whom I was meeting often as president of the Board of Directors of the European Journal "Astronomy and Astrophysics". Dr. Beiglböck constantly reminded me of my duty to write such a book. Finally I promised to him to do that. Then he hastened to send me a contract without waiting, first, for the book to be written.

The writing took about two years, but finally the book appeared in August 2002.

Before this book there had been no book on chaos in dynamical astronomy. Only the monumental 5-volume "Celestial Mechanics" of Y. Hagihara (1970) had some descriptions of the work of Hénon, and of others, on chaos. But there was not a special chapter on chaos.

On the other hand there have been several books on chaos in general, but not related to dynamical astronomy, or astronomy and astrophysics. I should mention two quite comprehensive books, one by Lichtenberg and Lieberman (1992) and the other by Gutzwiller (1990). More books, and many representative papers on order and chaos are mentioned in the references of my book.

My book deals with order and chaos from the perspective of dynamical astronomy. I included many special topics, which are of interest both to astronomers and to physicists in general.

There are already three published reviews of this book. The first, in Celestial Mechanics and Dynamical Astronomy 2003 says: "In this compendium... G. Contopoulos is mainly using his own vast experience, which he has after more than 50 years of research in this field. From the very beginning in the early sixties - the first computation of the third integral ... by the author of the book - up to modern work on the destruction of tori in connection to orbits captured in the stickiness regions... the reader can learn the basic ideas of chaos... For advanced students it is an excellent textbook concerning different aspects of dynamical systems... This outstanding book... closes a gap between celestial mechanics and stellar dynamics... it is of great interest for any physicists dealing with dynamical systems in general".

The second review is published in French in the "Courier des Astronomes Français (2003). It describes the contents of the book and it concludes: "The increasing interest for the non-linear dynamics has produced the publication of a large number of monographs these last years. But it was missing up to the present a synthetic work, where the reader could find the most recent developments in this domain and at the same time references to many papers necessary for research. This book fills well this gap. Written by a person that for half a century followed closely and, in many cases, has himself contributed quite essentially in the development of the theory of chaos, this book will be particularly useful to people preparing their thesis and will become indispensable for every astronomical library".

The third review is published in Mathematical Reviews. After describing the contents of the book it concludes: "Researchers in Dynamical Systems and in Astronomy and Astrophysics will find in this book an excellent source of information, covering the present understanding of the role of order and chaos in dynamics. Advanced students working in Dynamical Systems may use it as a complete textbook, where the complex interplay between order and chaotic motion is accurately illustrated in a systematic manner. Students in Dynamical Astronomy will appreciate in particular the two sections devoted to the dynamics of Galaxies and to the other astronomical applications. Finally, it may be used as a complementary text for students working not just in Dynamical Astronomy, but in any field for which non-linear dynamical phenomena are relevant".

My book found a good response in the astronomical community. The first edition has been sold out, after only 1 1/2 year.

56. PROSPECTS FOR THE FUTURE

The short range development of dynamical astronomy can be reasonably predicted. In my 2002 book on "Order and Chaos in Dynamical Astronomy" I give a list of open problems that can be attacked at the present time. These problems are not simple exercises, but neither are they deep unsolved problems, where progress is uncertain. Therefore, some good graduate students should be able to do useful research on them.

To find a good research problem is sometimes a matter of luck, or intuition. But "good" does not mean necessarily "easy". On the contrary one usually finds unexpected difficulties when dealing with a new subject, however interesting and straightforward.

In this connection I may mention the subject of chaos in relativity (section 36). When I started working in this field (1989) very little work had been done on this subject before. After the first example of chaos in the case of two fixed black holes (Contopoulos 1990), several papers extended this work to other cases. Nowadays the subject of chaos in relativity is a wide open field of research.

There is no doubt that the general subject of chaos will be prominent in many disciplines in the coming years. It is remarkable that celestial mechanics, that in the past dealt almost exclusively with order, now emphasizes more and more strongly chaos. Chaos in the three-body problem, chaos in the solar system, chaos in extrasolar systems, and so on.

The same development seems to start in the area of elementary particle physics. In the past physicists were looking for symmetries and integrability that led to important predictions and discoveries of new elementary particles and of new theories (supersymmetry, grand unified theories, and theories of everything). Now people start to realize that the symmetries are not perfect, that nonintegrability should play an important role in the most modern theories.

I expect that the theory of chaos will play an important role in our understanding of the universe and of its fundamental constituents.

Another aspect that comes to the forefront is the observational manifestation of chaos. Chaos can be established now by observations of the solar system and, more recently, by observations of galaxies. Such observations will be extended considerably in the years to come.

A new subject that has emerged in recent years refers to the role of nonintegrability versus integrability in partial differential equations (PDEs). Integrable PDEs have as special solutions the well known solitons and breathers. The main topic now is how these solitons change and disappear when the nonintegrability increases.

A more recent development in this field is the relation between integrable PDEs and integrable systems of ordinary differential equations (ODEs) (Voglis 2003a,b, etc.). This subject is presently under development and there is no doubt that it will be extended in the coming years.

Up to now most of the work on dynamical systems has been done in two degrees of freedom. Many aspects of these systems (but not all) are well understood today. The basic element of this study has been the theory of orbits.

On the other hand extensive work has been done also in the other extreme case, i.e. in systems with N degrees of freedom when N tends to infinity. In this case statistical methods play a most important role.

What is not well explored is the case of N degrees of freedom with N relatively small. The most important problem is to understand how the results from systems with a few degrees of freedom can be extended to many degrees of freedom. One example is the Fermi-Pasta-Ulam problem of N coupled oscillators. We know that for energies E below a critical energy E_{cr} there is no equipartition of the energy among the N modes. On the other hand for $E > E_{cr}$ equipartition is realized very fast. The question is what is the dependence of E_{cr} on the number N of oscillators. There are indications that the critical energy per particle E_{cr}/N is almost constant for a long time (Galgani and Scotti 1972; Galgani and Lo Vecchio 1979). However, because of Arnold diffusion (section 28) after a very long time equipartition will be realized. But this very long time may be longer than the age of the universe. Galgani et al. (1992) have shown that the equipartition time T_{eq} is reduced by a factor $\log N$ as N increases. Therefore T_{eq} tends to zero when $N \to \infty$. But even for N as large as the number of air molecules in a room (N of order 10^{23}) $\log N$ is only 23.

i.e. the reduction of T_{eq} is not so important. It seems that in many cases the thermodynamic limit is not realized in real dynamical systems.

Another strange result referring to large N-body systems was emphasized by Kandrup (2003). The Lyapunov characteristic number (section 28) increases with the number N of degrees of freedom. Therefore one expects the systems with large N to be very chaotic. However, large N-body systems usually have a quite regular overall form, e.g. close to spherical, and chaos is much less important. For this reason Kandrup (2003) made a distinction between two types of chaos. Microchaos, due to the discreteness of the N-bodies, and macrochaos referring to the bulk potential. Correspondingly he defined two Lyapunov characteristic numbers, that both appear in numerical simulations.

There are many problems in systems of N degrees of freedom that have not been attacked yet, or have been attacked only partly. Unexpected phenomena have been discovered in N-body systems, like the gravothermal catastrophe of Lynden-Bell and Wood (1968).

An apparently easier problem is the general three-body problem. But the study of the orbits in the three-body problem has shown an almost unbelievable complexity. Even the much simpler Sitnikov problem (a 3-D elliptical restricted 3-body problem, in which the 3rd (infinitesimal) body moves perpendicularly to the plane of motion of the other two bodies, through their center of mass), has provided unexpected complexities. For example there are orbits that make unbounded oscillations, without escaping to infinity (Sitnikov 1961).

These examples show that we are very far from understanding the general characteristics of systems with three or more degrees of freedom. It seems almost hopeless to try to derive the detailed dynamics of complicated systems with large N, like the DNA chains from first principles. However, such systems are formed in nature and they have some very remarkable stability properties.

Once we know from experiments the form of such complicated molecules, we can study their detailed structure, their oscillations, and their stability or instability. But it seems impossible to find ab initio such configurations with the help of computers. The computers needed for such a task would be much larger than the whole universe.

Another problem with possibly great consequences in physics is the development of a nonlinear quantum mechanical theory, as advocated by Penrose (1976, 1992) and Hawking and Penrose (1996). According to Penrose such a theory could lead to a supergrand unification in physics, namely the unification of gravity with the other forces of nature. Work in this direction is very limited nowadays, but there is no doubt that it will increase in the future. In fact, quantum mechanics is the only theory that is strictly linear and this leads to many difficulties and paradoxes. It is therefore natural to try to generalize it and check the validity of any such a generalization.

The paradoxes of the quantum theory are always a challenge for modern physics. In particular the Einstein-Podolsky-Rosen paradox leads to nonlocal phenomena that are a real puzzle nowadays. The experimental verification of Bell's inequalities (Bell 1964,1971, Aspect et al. 1982) indicate that nonlocal phenomena do occur. Therefore it is of interest to study such nonlocal phenomena in classical physics also.

A particular case of such a problem is the Abraham-Lorentz-Dirac equation, which contains not only second derivatives with respect to the time but also a third derivative. One may think that determinism is not affected by this. A motion should be determined if one gives the initial position, velocity and acceleration. However most solutions give runaway results that may be considered unphysical. Only if the initial acceleration has specific values the solution is finite. The restriction comes from a condition not at the origin, but at infinity, i.e. it is nonlocal.

Recent explorations of the Abraham-Lorentz-Dirac equation have been made by Galgani and his associates (Carati and Galgani 1999). Further explorations of such problems are highly desirable.

A new field of applications of dynamical theory is biophysics. The success in the exploration of the human genome is now followed by an exploration of the physico-chemical actions of the biological molecules (genes, and the constituents of cells of various types, including the brain cells). There is no doubt that extensive research will be devoted to this field for centuries to come.

It is already remarkable that many bright students turn now more towards biology rather than to the two most exciting subjects in physics, elementary particles and cosmology. But all these subjects need a sound background of mathematics and physics, including, in particular, dynamics. This will be always a necessary prerequisite in order to attack the more complicated modern problems of biology and biophysics. The future is wide open for such applications.

Thus, I conclude my book with a note of optimism. Science is not approaching its end, as claimed by a number of prominent scientists. It is open to unlimited possibilities that may bring unexpected discoveries in the years to come.

Acknowledgements

I would like to thank Drs. N. Voglis and C. Efthymiopoulos for reading a draft of this book and for their suggestions. I thank also Mrs. A. Zografaki for typing the manuscript and Mr. M. Zoulias for much technical help.

References

(In alphabetical and chronological order for one name and two names, and only in chronological order for more than two names)

Aspect, A., Dalibard, J. and Roger, G.: 1982, *Phys. Rev. Lett.* 49, 1804.

Bak, P.: 1997, *New Scientist*, 6th September 1997, 42.

Barbanis, B.: 1962, *Z.Astrophys.* 56, 56, and Thesis (in Greek).

Bartlett, J.H.: 1987, in Buchler, J.R. and Eichorn, H. (eds), "Chaotic Phenomena in Astrophysics", *New York Acad.Sci.Annals* 497, 78.

Bartlett, J.H.: 1989, *Cel. Mech.* 46, 129.

Belinskii, V. A. and Khalatnikov, I.M.:1969, *Sov.Phys.JETP* 29, 911.

Bell, J.S.: 1964, *Physics* 1,195.

Bell, J.S.: 1971, Proc. Intern. School Phys. "Fermi", Academic Press, New York, 33.

Benettin, G., Cercignani, C., Galgani, L. and Giorgilli, A.: 1980, *Lett.Nuovo Cim.* 28,1.

Binney, J. and Spergel, D.: 1982, *Astrophys. J.* 252, 308.

Birkhoff, G.D.: 1927, *"Dynamical Systems"*, Amer.Math.Soc., Providence, R.I.

Birkhoff, G.D.: 1935, *Mem.Pontif.Acad.Sci.Novi Lyncaei* s.3, 1,85.

Blaauw, A.: 1991, *"ESO's Early History"*, ESO.

Bozis, G.: 1966, *Astron. J.* 71, 404.

Bozis, G.: 1967, *Astron. J.* 72, 380.

Brack, M.: 2001, *Found.Physics* 31, 209.

Broucke, R.: 1969, *Am.Inst.Aeronaut.Astronaut.J.* 7, 1003.

Brouwer, D. and Clemence, G.M.: 1961, *"Celestial Mechanics"*, Academic Press, New York.

Buchler, J.R., Gottesman, S.T. and Kandrup, H.E. (eds): 1998, "Nonlinear Dynamics and Chaos in Astrophysics. A Festschrift in Honor of George Contopoulos", *Ann. New York Acad. Sci.* 867.

Candlestickmaker, S.: 1972, *Quart. J.Roy. Astr.Soc.* 13,63.

Carati, A. and Galgani, L.: 1999, *Nuovo Cim.B* 114, 489.

Cercignani, C.: 1977, *Riv. Nuovo Cim.* 7,429.

Chandrasekhar, S.: 1939, *Astrophys. J.* 90, 1.

Chandrasekhar, S.: 1940, *Astrophys. J.* 92, 441.

Chandrasekhar, S.: 1942, *"Principles of Stellar Dynamics"*, Univ. Chicago Press, Chicago; Dover Publ. (1960).

Chandrasekhar, S.: 1983, *"The Mathematical Theory of Black Holes"*, Clarendon Press.

Chandrasekhar, S.: 1989, *Proc.Roy.Soc.London A* 421, 227.

Chandrasekhar, S.: 1995, *"Newton's Principia for the common reader"*, Oxford, Clarendon Press.
Chandrasekhar, S. and Contopoulos, G.: 1963, *Proc.Nat.Acad.Sci.* 49, 608.
Chandrasekhar, S. and Contopoulos, G.: 1967, *Proc.Roy.Soc.* A 298, 23.
Chandrasekhar, S. and Nutku,Y.: 1969, *Astrophys. J.* 158, 55.
Cherry, T.M.,: 1924a, *Proc. Cambridge Phil.Soc.* 22, 273, 287, 325 and 510.
Chirikov, B.V.: 1959, *Atomnaya Energiya* 6, 630.
Chirikov, B.V.: 1979, *Phys.Rep.* 52, 263.
Contopoulos, G.: 1954, *Z. Astrophys.* 35, 67.
Contopoulos, G.: 1956a, *Z. Astrophys.* 39, 126.
Contopoulos, G.: 1956b, *Astrophys. J.* 124, 643.
Contopoulos, G.: 1957a, *Z. Astrophys.* 42, 75.
Contopoulos, G.: 1957b, *Stockholm Ann.* 19, No 10.
Contopoulos, G.: 1958, *Stockholm Ann.* 20, No 5.
Contopoulos, G.: 1960, *Z. Astrophys.* 49, 273.
Contopoulos, G.: 1963a, *Astron. J.* 68, 763.
Contopoulos, G.: 1963b, *Astron. J.* 68, 1.
Contopoulos, G.: 1965a, *Astron. J.* 70, 526.
Contopoulos, G.: 1965b, *Astrophys. J.* 142, 802.
Contopoulos, G.: 1966a, "Problems of Stellar Dynamics", in "Space Mathematics I", Amer. Math. Soc. 169.
Contopoulos, G.: 1966b, *Astrophys. J.* Suppl. 13, 503.
Contopoulos, G. (ed): 1966c, "The Theory of Orbits in the Solar System and in Stellar Systems", IAU Symp. 25, Academic Press, London.
Contopoulos, G. (ed): 1966d, in Hénon M. and Nahon, F. (eds), "Les Nouvelles Méthodes de la Dynamique Stellaire"; *Bull.Astron.* (3) 2, 223 (1967).
Contopoulos, G.: 1967a, *Astron. J.* 72, 669.
Contopoulos, G.: 1967b, *Astron. J.* 72,191.
Contopoulos, G.: 1967c, Proc.14th Liège Symp. "Gravitational Instability and Formation of Stars and Galactic Structure", 213.
Contopoulos, G.: 1968, *Astrophys. J.* 153, 83.
Contopoulos, G.: 1970a, *Astrophys. J.* 160,113.
Contopoulos, G.: 1970b, *Astron. J.* 75, 96.
Contopoulos, G.: 1970c, *Astron. J.* 75, 108.
Contopoulos, G.: 1971a, *Astrophys.J.* 163, 181.
Contopoulos, G.: 1971b, *Astron. J.* 76, 147.
Contopoulos, G.: 1972, "The Dynamics of Spiral Structure.Lecture Notes", Univ. of Maryland.
Contopoulos, G.: 1973a, in Martinet, L. and Mayor, M. (eds) "Dynamical Structure and Evolution of Stellar Systems", Geneva Observatory, 52.
Contopoulos, G.: 1973b, *Astrophys. J.* 181,657.
Contopoulos, G.: 1975a, *Astrophys. J.* 201,566.
Contopoulos, G.: 1975b, in Hayli, A.(ed.) "Dynamics of Stellar Systems", IAU Symp. 69,209.
Contopoulos, G.: 1978a, in Lebovitz, N.R, Reid, W.H. and Vandervoort, P.O. (eds) "Theoretical Principles in Astrophysics and Relativity, Univ. of Chicago Press, 93.
Contopoulos, G.: 1978b, *Cel.Mech.* 17, 167.
Contopoulos, G.: 1979, *Cel.Mech.* 18, 195.
Contopoulos, G.: 1980, *Astron.Astrophys.* 81, 198.
Contopoulos, G.: 1981a, *Astron.Astrophys.* 104, 156.
Contopoulos, G.: 1981b, *Lett. Nuovo Cim.* 30, 498.
Contopoulos, G.: 1981c, *Astron.Astrophys.* 102, 265.

Contopoulos, G.: 1983a, *Lett. Nuovo Cim.* 37, 149.
Contopoulos, G.: 1983b, *Astron.Astrophys.* 117, 89.
Contopoulos, G.: 1985, *Comments Astrophys.* 11,1.
Contopoulos, G.: 1986a, *Cel. Mech.* 38, 1.
Contopoulos, G.: 1986b, *Astron.Astrophys.* 161,244.
Contopoulos, G.: 1986c, *Particle Accelerators* 19,107.
Contopoulos, G.: 1988a, *Cel. Mech.* 42, 239.
Contopoulos, G.: 1988b, *Cel. Mech.* 44, 393.
Contopoulos, G.: 1988c, *Astron.Astrophys.* 201, 44.
Contopoulos, G. : 1988d, in Valtonen, M.J. (ed) "The Few-Boy Problem", IAU Coll. 96, Kluwer, Dordrecht, 265.
Contopoulos, G.: 1990a, *Astron.Astrophys.* 231, 41.
Contopoulos, G.: 1990b, *Proc.Roy.Soc.London A* 431, 183.
Contopoulos, G.: 1991, *Proc.Roy.Soc.London A* 435, 551.
Contopoulos, G.: 1993 in Buchler, J.R. and Kandrup, H. (eds) "Stochastic Processes in Astrophysics", *New York Acad. Sci. Ann.* 706, 100.
Contopoulos, G.: 1994, in Contopoulos, G., Spyrou, N.K. and Vlahos, L. (eds) "Galactic Dynamics and N-Body Simulations", Springer Verlag, New York, 33.
Contopoulos, G.: 2003, "Order and Chaos in Dynamical Astronomy", SpringerVerlag, New York, 3.
Contopoulos, G. and Barbanis, B.: 1968, *Astroph.Space Sci.* 2, 134.
Contopoulos, G. and Barbanis, B.: 1985, *Astron. Astrophys.* 153, 44.
Contopoulos, G. and Bozis, G.: 1964, *Astrophys.J.* 139, 1239.
Contopoulos, G. and Doukoumetzidis,A.: 2004, in Laskarides, P.G.(ed.), Proc.6th Hellenic Astron. Conference, 267.
Contopoulos, G. and Efstathiou, C.: 2004, *Cel.Mech.Dyn.Astron.* 88, 163.
Contopoulos, G. and Giorgilli, A.: 1988, *Meccanica* 23, 19.
Contopoulos, G. and Grosbøl, P.: 1988, *Astron. Astrophys.* 197, 83.
Contopoulos, G. and Grosbøl, P.: 1989, *Astron. Astrophys. Rev.* 1, 261.
Contopoulos, G. and Harsoula, M.: 2004, *J.Math.Phys.* (in press)
Contopoulos, G. and Kaufmann, D.: 1992, *Astron.Astrophys.* 253, 379.
Contopoulos, G. and Kotsakis, D.: 1986 "Cosmology. The Structure and Evolution of the Universe", Springer Verlag, New York.
Contopoulos, G. and Magnenat, P.: 1985, *Cel. Mech.* 37, 387.
Contopoulos, G. and Mertzanides, C.: 1977, *Astron. Astrophys.* 61, 477.
Contopoulos, G. and Moutsoulas, M.: 1965, *Astron.J.* 70, 817.
Contopoulos, G. and Moutsoulas, M.: 1966, *Astron.J.* 71, 687.
Contopoulos, G. and Polymilis, C.: 1987, *Physica D* 24, 328.
Contopoulos, G. and Polymilis, C.: 1993, *Phys. Rev. E* 47, 1546.
Contopoulos, G. and Polymilis, C.: 1996, *Cel.Mech.Dyn.Astron.* 63, 189.
Contopoulos, G. and Seimenis, J.: 1990, *Astron.Astrophys.* 227, 49.
Contopoulos, G. and Strömgren, B.: 1965, "Tables of Plane Galactic Orbits", NASA Inst. Space Studies, New York.
Contopoulos, G. and Vandervoort, P.: 1992, *Astrophys. J.* 389, 118.
Contopoulos G. and Vlahos, L.: 1975, *J. Math. Phys.* 16, 1469.
Contopoulos G. and Voglis, N.: 1996, *Cel. Mech.Dyn.Astron.* 64, 1.
Contopoulos G. and Voglis, N.: 2003, (eds) "Galaxies and Chaos", Springer Verlag, Heidelberg
Contopoulos G. and Woltjer, L.: 1964, *Astrophys. J.* 140, 1106.
Contopoulos, G. and Zikides, M.: 1980, *Astron. Astrophys.* 90, 198.
Contopoulos, G., Galgani, L. and Giorgilli, A.: 1978, *Phys. Rev. A* 18, 1183.

Contopoulos, G., Gottesman, S.T., Hunter, J.H. Jr. and England, M.N:1989, *Astrophys.J.* 343, 608.
Contopoulos, G., Grammaticos, B. and Ramani, A.: 1993, *J. Phys.A* 26, 5795.
Contopoulos, G., Papadaki, H. and Polymilis, C.: 1994a, *Cel.Mech.Dyn.Astron.* 60,249.
Contopoulos, G., Spyrou, N.K. and Vlahos, L. (eds): 1994b, "Galactic Dynamics and N-Body Simulations", Springer Verlag, Heidelberg.
Contopoulos, G., Farantos, S.C., Papadaki, H. and Polymilis, C.: 1994c, *Phys.Rev. E* 50, 4399.
Contopoulos, G., Papadaki, H. and Polymilis, C.: 1995a, in Roy, A.E. (ed.), "From Newton to Chaos" Plenum Press, 485.
Contopoulos, G. Grammaticos, B. and Ramani, A.: 1995b, *J. Phys.A* 28, 5313.
Contopoulos, G., Voglis, N., Efthymiopoulos, C. and Grousousakou, E.: 1995c, in Hunter, J. and Wilson, R (eds) "Waves in Astrophysics", *New York Acad. Sci.Annals* 723,145.
Contopoulos, G., Voglis, N. and Efthymiopoulos, C.: 1996, Nobel Symposium, 98,19.
Contopoulos, G., Voglis, N., Efthymiopoulos, C., Froeschlé, C., Gonczi,R., Lega, E.,Dvorak, R. and Lohinger, E.: 1997, *Cel.Mech.Dyn.Astron.* 67, 293.
Contopoulos, G., Voglis, N. and Efthymiopoulos, C.: 1999a, *Cel.Mech.Dyn.Astron.* 73, 1.
Contopoulos, G., Harsoula, M., Voglis, N. and Dvorak, R.: 1999b, *J. Phys.A* 22, 5213.
Contopoulos, G., Efthymiopoulos, C. and Voglis, N.: 2000, *Cel.Mech.Dyn.Astron.* 78, 243.
Contopoulos, G., Voglis, N. and Kalapotharakos, C.: 2002, *Cel.Mech.Dyn.Astron.* 83, 191.
Contopoulos, G., Dvorak, R., Harsoula M. and Freistetter, F.: 2004, *Bifurcation and Chaos* (in press).
Cornish, N.J. and Levin, J.J.: 1997, Phys.Rev.Lett. 78, 998; *Phys.Rev.D* 55,7489, 1997.
Coullet, P. and Tresser, C.: 1978, *J. de Physique* 39, C5, 25
Cushman, R. and Sniatycki, J.: 1995, *Rep.Math.Phys.* 36, 75.
Davis, M.J. and Heller, E.J.: 1981, *J. Chem.Phys.* 75, 246.
de Zeeuw, T.:1985, *Mon.Not.R.Astr.Soc.* 216, 273.
de Zeeuw, T., Hunter, C. and Schwarzschild, M.:1987, *Astrophys. J.* 317, 607.
Dragt, A. and Finn, J.M.: 1976, *J.Geophys.Res.* 81, 2327.
Dragt, A. and Finn, J.M.: 1979, *J.Math.Phys.* 20, 2649.
Dragt, A. et al.: 1988, *Ann. Rev. Nucl. Part. Sci.* 38, 455.
Duarte, P.: 1994, *Ann.Inst.Henri Poincaré* 11, 359.
Dvorak,R., Funk, B., Freistetter, F and Contopoulos, G.: 2003, in Freistetter, F., Dvorak, R. and Erdi, B.(eds) "3rd Austrian-Hungarian Workshop on Trojans and Related topics", Eötvös Univ.Press, Budapest, 185.
Eckhardt, B., Ford, J. and Vivaldi, F.: 1984, *Physica D* 13,339.
Efstathiou, K. and Contopoulos, G.: 2001, *Chaos* 11, 327.
Efthymiopoulos, C.: 2004, Proc. 6th Humbold Symposium (in press)
Efthymiopoulos, C., Contopoulos, G., Voglis, N. and Dvorak, R.: 1997, *J.Phys. A* 30, 5167.
Efthymiopoulos, C., Giorgilli, A. and Contopoulos, G.: 2004, *J.Phys. A* (in press).
Einstein, A. and Stern, O.: 1913, *Ann.Physik* 40, 551.
Farantos, S.C., Founariotakis, M. and Polymilis, C.: 1989, *Chem.Phys.* 135, 347.
Feigenbaum, M.: 1978, *J. Stat. Phys.* 19, 25.
Fermi, E., Pasta, J. and Ulam, S.: 1955, Los Alamos Lab.Rep. LA 1949.
Flashka, H: 1974, *Phys. Rev. B* 9, 1924.
Ford, J.: 1975, in Cohen, E.D.G.(ed) "Fundamental Problems in Statistical Mechanics", North Holland, Amsterdam, 3, 215.
Ford, J., Stoddard, S.D. abd Turner, J.S.: 1973, *Prog.Theor.Phys.*, 50, 1547.
Ford, J., Mantica, G. and Ristow, G.H.: 1991, *Physica D* 50,493.
Founariotakis, M., Farantos, S.C., Contopoulos, G. and Polymilis, C.: 1989, *J.Chem.Phys.* 91,1389.

Freeman, K.C.: 1966, *Mon.Not.R.Astr.Soc.* 133, 47; 134, 1 and 15.
Froeschlé, C.: 1972, *Astron.Astrophys.* 16, 172.
Froeschlé, C., Froeschlé, Ch. and Lohinger, E.: 1993, *Cel.Mech.Dyn.Astron.* 56,307.
Fujimoto, M.: 1968, *Astrophys. J.* 152, 391.
Galgani, L. and Lo Vecchio, G.: 1979, *Nuovo Cim. B* 55,1.
Galgani, L. and Scotti, A.: 1972, *Riv.Nuovo Cim.* 2,189.
Galgani, L., Giorgilli, A., Martineli, A. and Vanzini, S.: 1992, *Physica D* 59,334.
Giorgilli, A.: 1979, *Computer Phys. Comm.* 16, 331.
Giorgilli, A. and Galgani, L.: 1978, *Cel. Mech.* 17, 267.
Giorgilli, A. and Skokos, Ch.: 1997, *Astron. Astrophys.* 317, 254.
Giorgilli, A., Delshams, A., Fontich, E., Galgani, L and Simó, C.: 1989, *J. Diff. Equ.* 77,167.
Grossman, S. and Thomae, S.: 1977, *Z. Naturforschung A* 32, 1353.
Guckenheimer, J. and Holmes, P.: 1983, "Nonlinear Oscillations, Dynamical Systems and Bifurcations of Vector Fields", Springer Verlag, New York.
Gurzadyan, V.G. and Savvidy, G.K.: 1986, *Astron.Astrophys.* 160, 203.
Gustavson, F.G.: 1966, *Astron. J.* 71, 670.
Guth, A.H.: 1981, *Phys.Rev. D* 23, 347.
Guth, A.H.: 1998, "The Inflationary Universe: The Quest for a New Theory of Cosmic Origins", Perseus Group Books.
Gutzwiller, M.C.: 1990, "Chaos in Classical and Quantum Mechanics", SpringerVerlag, New York.
Hadjidemetriou, J. D.:1975, *Cel. Mech.* 12,255.
Hagihara, Y.: 1970, "Celestial Mechanics", 1; 2(1 and 2) (1972) MIT Press, Cambridge; 3(1 and 2) (1974), 4(1975) Japan Society for the Promotion of Science, Tokyo.
Hawking, S. and Penrose, R.: 1996, *"The Nature of Space and Time"*, Princeton Univ. Press.
Heggie, D.C.: 1983, *Cel. Mech.* 29, 207.
Heller, E.J.: 1993, in Giannoni et al. (eds), *"Chaos in Quantum Physics"*, Elsevier.
Hénon, M.: 1960, *Ann.Astrophys.* 23,668.
Hénon, M.: 1966, *Bull. Astron.* (3) 1(2), 49.
Hénon, M.: 1974, *Phys. Rev. B* 9, 1921.
Hénon, M. and Heiles, C.: 1964, *Astron. J.* 69, 73.
Hietarinta, J.: 1987, *Phys. Rep.* 147, 87.
Hobill, D., Burd, A. and Coley, A. (eds):1994, *"Deterministic Chaos in General Relativity"*, Plenum Press, New York.
Hori, G.I.: 1962, *Publ. Astron. Soc. Japan* 14, 353.
Hunter, C.: 1990 in Buchler, J.R., Gottesman, S.T. and Hunter, J.H. Jr. (eds) *"Galactic Models"*, New York Acad.Sci.Ann. 596, 187.
Kandrup, H.E.: 2003, in Contopoulos, G. and Voglis, N. (eds), *"Galaxies and Chaos"*, Springer Verlag, New York, 154.
Kandrup, H.E., Siopis, C., Contopoulos, G. and Dvorak, R.: 1999, *Chaos* 9, 381.
Karney, C.F.F.: 1983, *Physica D* 8,387.
Kazanas, D.: 1980, *Astrophys. J. Lett.* 241, L59.
Krefetz, E.: 1967, *Astron. J.* 72, 471.
Kurth, R: 1955, *Astron. Nachr.* 282, 97.
Kurth, R: 1957, *"Introduction to the Dynamics of Stellar Systems"*,Pergamon Press, New York.
Landau, L.D. and Lifshitz, E.M.: 1960, *"Mechanics"*, Pergamon Press, New York (1st edition).
Latifi, A., Musette, M. and Conte, R.: 1994, *Phys. Lett. A* 194, 83.
Lecar. M.: 1968, *Bull. Astron.* 3, 91.
Lichtenberg, A.J and Lieberman, M.A.: 1992, *"Regular and Chaotic Dynamics"*, 2nd Ed., Springer Verlag, New York.

Lovelace, R.V.E. and Hohlfeld, R.G.: 1978, *Astrophys. J.* 221, 51.
Lynden-Bell, D. : 1998, in Buchler, J.R., Gottesman, S.T. and Kandrup, H.E. (eds) "Nonlinear Dynamics and Chaos in Astrophysics", *New York Acad. Sci. Ann.* 867, 3.
Lynden-Bell, D. and Wood, R.: 1968, *Mon.Not.R.Astr.Soc.* 138, 495.
Mandelbrot, B.B.: 1982, *"The Fractal Geometry of Nature"*, Freeman, San Francisco.
Merritt, D.: 1999, *Proc.Astr.Soc.Pacific* 111, 129.
Misner, C.M.: 1969, *Phys.Rev.Lett.* 22, 1071.
Miura, R.M., Gardner, C.S. and Kruskal, M.D.: 1968, *J.Math.Phys.* 9,1204.
Morbidelli, A. and Giorgilli, A.: 1995, *J. Stat. Phys.* 78, 1607.
Moser, J.: 1968, *Mem. Amer. Math. Soc.* 81, 1.
Nekhoroshev, N.N.: 1977, *Russ. Math. Surv.* 32(6), 1.
Newhouse, S.E.: 1983, in Iooss, G., Helleman, R.H.G. and Stora, R. (eds) *"Chaotic Behaviour of Deterministic Systems"*, North Holland, Amsterdam, 443.
Ossipkov, L.P.: 1977, *Vestnik Lenigrad Univ.* 140.
Patsis, P.A., Contopoulos, G. and Grosbøl, P.: 1991, *Astron. Astrophys.* 243, 373.
Patsis, P.A.,Hiotelis, N., Contopoulos, G. and Grosbøl, P.: 1994, *Astron. Astrophys.* 286,46.
Penrose, R.: 1976, *Gen.Rel.Grav.* 7, 31.
Penrose, R.: 1992, in Zichichi, A., de Sabbata, N. and Sanchez, N (eds) "Gravitation and Modern Cosmology", Plenum Press, New York.
Perdang, J. and Blacher, S.: 1982, Astron.Astrophys. 112, 35.
Perdang, J. and Blacher, S.: 1984, *Astron.Astrophys.* 136, 263.
Pfenniger, D.: 1984, *Astron.Astrophys.* 141,171.
Poincaré, H.: 1892, *"Les Méthodes Nouvelles de la Méchanique Céleste"* Gauthier Villars, Paris, I (1892), II (1893), III (1899); Dover edition (1957).
Prendergast, K.: 1982, in Chudnovsky, D. and G (eds) *"The Riemann Problem"*, Springer Verlag, New York.
Prigogine, I.: 1997 *"The End of Certainty"*, Free Press.
Rosenbluth, M.N., Sagdeev, R. A., Taylor, J. B. and Zaslavsky, G.M.: 1966, *Nucl. Fusion* 6, 217.
Rowan-Robinson, M.: 1988, *Nature* 332, 188.
Schmutzer, E.: 1989, *Gen. Relativity and Gravitation* 21, 1.
Schwarzschild, M.: 1979, *Astrophys. J.* 232, 236.
Scott, A.C., Chu, F.Y.F. and McLaughlin, D.W.: 1973, *Proc. IEEE* 61, 1443.
Sellwood, J.A. and Lin, D.N.C: 1989, *Mon.Not.R.Astr.Soc.* 240, 991.
Sellwood, J.A. and Sparke, L.S.: 1988, *Mon.Not.R.Astr.Soc.* 231, 25.
Shirts, R.B. and Reinhardt, W.P.: 1982, *J. Chem. Phys.* 77, 5204.
Sitnikov, K.: 1961, *Soviet Phys. Dokl.* 5, 647.
Skokos, C., Contopoulos, G. and Giorgilli, A.: 1996, Proc. 2nd Hellenic Astron. Conf. 526.
Smale, S.: 1963, in Cairns, S.S. (ed) *"Differential and Combinatorial Topology"*, Princeton Univ. Press, Princeton, 63.
Smale, S.: 1967, *Bull.Am.Math.Soc.* 73, 747.
Strömgren, E.: 1917, *Astron. Nachr.* 203, 16.
Surdin, V.G.: 1988, *Sov. Astron.* 32, 345.
Toda, M.: 1970, *Prog. Theor. Phys. Suppl.* 45, 174.
Toomre, A. and Toomre, J.: 1972, *Astrophys. J.* 178, 623.
Torgard, I. and Ollongren, A.: 1960, *Nuffic Intern.Summer Course in Science, Part X.*
Udry, S. and Pfenniger, D.: 1988, *Astron. Astrophys.* 198, 135.
Uzer, T., Farrelly, D., Milligan, J.A., Raines, P.E. and Skelton, J.P.: 1991, *Science* 253,42.
Vandervoort, P.O.: 1971, *Astrophys. J.* 166, 37.
Vandervoort, P.O.: 1979, *Astrophys. J.* 232, 91.

Varvoglis, H., Tsiganis, K. and Hadjidemetriou, J.D.: 2003, in Contopoulos, G. and Voglis, N. (eds), *"Galaxies and Chaos"*,Springer Verlag, New York, 433.
Voglis, N.: 2003a, *Mon.Not.Roy.Astr.Soc.* 344, 578.
Voglis, N.: 2003b, in Contopoulos, G. and Voglis, N. (eds), *"Galaxies and Chaos"*, Springer Verlag, New York, 56.
Voglis, N. and Contopoulos, G.: 1994, *J. Phys.A* 27, 4899.
Voglis, N., Kalapotharakos, C. and Stavropoulos, I.: 2002, *Mon.Not.R.Astr.Soc.* 337, 619.
Whittaker, E.T.: 1916, *Proc. Roy. Soc. Edinburgh* 37, 95.
Whittaker, E.T. : 1937, *"A Treatise on the Analytical Dynamics of Particles and Rigid Bodies"*, 4th Ed., Cambridge Univ. Press, Cambridge.
Yoshida, H., Ramani, A. and Grammaticos, B.: 1988, *Physica D* 30, 151.
Zabusky, N.J. and Kruskal, M.D.: 1965, *Phys.Rev.Lett.* 15,240.
Zacharov, V.E. and Fadeev, L.D.: 1971, *Func.Anal.Appl.* 5, 280.
Zachilas, L.G.: 1993, *Astron. Astrophys. Suppl.* 97,549.
Zaslavsky, G.M.: 1995, *Chaos* 5, 653.
Zaslavsky, G.M.: 1999, *Physics Today*, August, 39.
Zaslavsky, G.M. and Chirikov, B.V.: 1972, *Sov.Phys.Uspekhi* 14, 549.
Zeldovich, Y.: 1983, *Highlights of Astronomy* 6, 32.

Astrophysics and Space Science Library

Volume 316: ***Civic Astronomy - Albany's Dudley Observatory, 1852-2002,*** by G. Wise
Hardbound ISBN 1-4020-2677-3, October 2004

Volume 315: ***How does the Galaxy Work - A Galactic Tertulia with Don Cox and Ron Reynolds,*** edited by E. J. Alfaro, E. Pérez, J. Franco
Hardbound ISBN 1-4020-2619-6, September 2004

Volume 314: ***Solar and Space Weather Radiophysics - Current Status and Future Developments,*** edited by D.E. Gary and C.U. Keller
Hardbound ISBN 1-4020-2813-X, August 2004

Volume 313: ***Adventures in Order and Chaos – A Scientific Autobiography,*** by G. Contopoulos
Hardbound ISBN 1-4020-3039-8, December 2004

Volume 312: ***High-Velocity Clouds,*** edited by H. van Woerden, U. Schwarz, B. Wakker
Hardbound ISBN 1-4020-2813-X, September 2004

Volume 311: ***The New ROSETTA Targets- Observations, Simulations and Instrument Performances***, edited by L. Colangeli, E. Mazzotta Epifani, P. Palumbo
Hardbound ISBN 1-4020-2572-6, September 2004

Volume 310: ***Organizations and Strategies in Astronomy 5,*** edited by A. Heck
Hardbound ISBN 1-4020-2570-X, September 2004

Volume 309: ***Soft X-ray Emission from Clusters of Galaxies and Related Phenomena***, edited by R. Lieu and J. Mittaz
Hardbound ISBN 1-4020-2563-7, September 2004

Volume 308: ***Supermassive Black Holes in the Distant Universe,*** edited by A.J. Barger
Hardbound ISBN 1-4020-2470-3, August 2004

Volume 307: ***Polarization in Spectral Lines***, by E. Landi Degl'Innocenti and M. Landolfi
Hardbound ISBN 1-4020-2414-2, August 2004

Volume 306: ***Polytropes – Applications in Astrophysics and Related Fields,*** by G.P. Horedt
Hardbound ISBN 1-4020-2350-2, September 2004

Volume 305: ***Astrobiology: Future Perspectives,*** edited by P. Ehrenfreund, W.M. Irvine, T. Owen, L. Becker, J. Blank, J.R. Brucato, L. Colangeli, S. Derenne, A. Dutrey, D. Despois, A. Lazcano, F. Robert
Hardbound ISBN 1-4020-2304-9, July 2004
Paperback ISBN 1-4020-2587-4, July 2004

Volume 304: ***Cosmic Gammy-ray Sources,*** edited by K.S. Cheng and G.E. Romero
Hardbound ISBN 1-4020-2255-7, September 2004

Volume 303: ***Cosmic rays in the Earth's Atmosphere and Underground***, by L.I, Dorman
Hardbound ISBN 1-4020-2071-6, August 2004

Volume 302: ***Stellar Collapse,*** edited by Chris L. Fryer
Hardbound, ISBN 1-4020-1992-0, April 2004

Volume 301: ***Multiwavelength Cosmology***, edited by Manolis Plionis
Hardbound, ISBN 1-4020-1971-8, March 2004

Volume 300: ***Scientific Detectors for Astronomy,*** edited by Paola Amico, James W. Beletic, Jenna E. Beletic
Hardbound, ISBN 1-4020-1788-X, February 2004

Volume 299: ***Open Issues in Local Star Fomation,*** edited by Jacques Lépine, Jane Gregorio-Hetem
Hardbound, ISBN 1-4020-1755-3, December 2003

Volume 298: ***Stellar Astrophysics - A Tribute to Helmut A. Abt,*** edited by K.S. Cheng, Kam Ching Leung, T.P. Li
Hardbound, ISBN 1-4020-1683-2, November 2003

Volume 297: ***Radiation Hazard in Space,*** by Leonty I. Miroshnichenko
Hardbound, ISBN 1-4020-1538-0, September 2003

Volume 296: ***Organizations and Strategies in Astronomy, volume 4,*** edited by André Heck
Hardbound, ISBN 1-4020-1526-7, October 2003

Volume 295: ***Integrable Problems of Celestial Mechanics in Spaces of Constant Curvature,*** by T.G. Vozmischeva
Hardbound, ISBN 1-4020-1521-6, October 2003

Volume 294: ***An Introduction to Plasma Astrophysics and Magnetohydrodynamics,*** by Marcel Goossens
Hardbound, ISBN 1-4020-1429-5, August 2003
Paperback, ISBN 1-4020-1433-3, August 2003

Volume 293: ***Physics of the Solar System,*** by Bruno Bertotti, Paolo Farinella, David Vokrouhlický
Hardbound, ISBN 1-4020-1428-7, August 2003
Paperback, ISBN 1-4020-1509-7, August 2003

Volume 292: ***Whatever Shines Should Be Observed,*** by Susan M.P. McKenna-Lawlor
Hardbound, ISBN 1-4020-1424-4, September 2003

Volume 291: ***Dynamical Systems and Cosmology***, by Alan Coley
Hardbound, ISBN 1-4020-1403-1, November 2003

Volume 290: ***Astronomy Communication,*** edited by André Heck, Claus Madsen
Hardbound, ISBN 1-4020-1345-0, July 2003

Volume 287/8/9: ***The Future of Small Telescopes in the New Millennium,*** edited by Terry D. Oswalt
Hardbound Set only of 3 volumes, ISBN 1-4020-0951-8, July 2003

Volume 286: ***Searching the Heavens and the Earth: The History of Jesuit Observatories,*** by Agustín Udías
Hardbound, ISBN 1-4020-1189-X, October 2003

Volume 285: ***Information Handling in Astronomy - Historical Vistas,*** edited by André Heck
Hardbound, ISBN 1-4020-1178-4, March 2003

Volume 284: ***Light Pollution: The Global View,*** edited by Hugo E. Schwarz
Hardbound, ISBN 1-4020-1174-1, April 2003

Volume 283: ***Mass-Losing Pulsating Stars and Their Circumstellar Matter,*** edited by Y. Nakada, M. Honma, M. Seki
Hardbound, ISBN 1-4020-1162-8, March 2003

Volume 282: ***Radio Recombination Lines,*** by M.A. Gordon, R.L. Sorochenko
Hardbound, ISBN 1-4020-1016-8, November 2002

Volume 281: ***The IGM/Galaxy Connection,*** edited by Jessica L. Rosenberg, Mary E. Putman
Hardbound, ISBN 1-4020-1289-6, April 2003

Volume 280: ***Organizations and Strategies in Astronomy III,*** edited by André Heck
Hardbound, ISBN 1-4020-0812-0, September 2002

Volume 279: ***Plasma Astrophysics , Second Edition,*** by Arnold O. Benz
Hardbound, ISBN 1-4020-0695-0, July 2002

Volume 278: ***Exploring the Secrets of the Aurora,*** by Syun-Ichi Akasofu
Hardbound, ISBN 1-4020-0685-3, August 2002

Volume 277: ***The Sun and Space Weather,*** by Arnold Hanslmeier
Hardbound, ISBN 1-4020-0684-5, July 2002

Volume 276: ***Modern Theoretical and Observational Cosmology,*** edited by Manolis Plionis, Spiros Cotsakis
Hardbound, ISBN 1-4020-0808-2, September 2002

Volume 275: ***History of Oriental Astronomy,*** edited by S.M. Razaullah Ansari
Hardbound, ISBN 1-4020-0657-8, December 2002

Volume 274: ***New Quests in Stellar Astrophysics: The Link Between Stars and Cosmology,*** edited by Miguel Chávez, Alessandro Bressan, Alberto Buzzoni,Divakara Mayya
Hardbound, ISBN 1-4020-0644-6, June 2002

Volume 273: ***Lunar Gravimetry,*** by Rune Floberghagen
Hardbound, ISBN 1-4020-0544-X, May 2002

Volume 272:***Merging Processes in Galaxy Clusters,*** edited by L. Feretti, I.M. Gioia, G. Giovannini
Hardbound, ISBN 1-4020-0531-8, May 2002

Volume 271: ***Astronomy-inspired Atomic and Molecular Physics,*** by A.R.P. Rau
Hardbound, ISBN 1-4020-0467-2, March 2002

Volume 270: ***Dayside and Polar Cap Aurora,*** by Per Even Sandholt, Herbert C. Carlson, Alv Egeland
Hardbound, ISBN 1-4020-0447-8, July 2002

Volume 269: ***Mechanics of Turbulence of Multicomponent Gases,*** by Mikhail Ya. Marov, Aleksander V. Kolesnichenko
Hardbound, ISBN 1-4020-0103-7, December 2001

Volume 268: ***Multielement System Design in Astronomy and Radio Science,*** by Lazarus E. Kopilovich, Leonid G. Sodin
Hardbound, ISBN 1-4020-0069-3, November 2001

Volume 267: ***The Nature of Unidentified Galactic High-Energy Gamma-Ray Sources,*** edited by Alberto Carramiñana, Olaf Reimer, David J. Thompson
Hardbound, ISBN 1-4020-0010-3, October 2001

Volume 266: ***Organizations and Strategies in Astronomy II,*** edited by André Heck
Hardbound, ISBN 0-7923-7172-0, October 2001

Volume 265: ***Post-AGB Objects as a Phase of Stellar Evolution,*** edited by R. Szczerba, S.K. Górny
Hardbound, ISBN 0-7923-7145-3, July 2001

Volume 264: ***The Influence of Binaries on Stellar Population Studies,*** edited by Dany Vanbeveren
Hardbound, ISBN 0-7923-7104-6, July 2001

Volume 262: ***Whistler Phenomena - Short Impulse Propagation,*** by Csaba Ferencz, Orsolya E. Ferencz, Dániel Hamar, János Lichtenberger
Hardbound, ISBN 0-7923-6995-5, June 2001

Volume 261: ***Collisional Processes in the Solar System,*** edited by Mikhail Ya. Marov, Hans Rickman
Hardbound, ISBN 0-7923-6946-7, May 2001

Volume 260: ***Solar Cosmic Rays,*** by Leonty I. Miroshnichenko
Hardbound, ISBN 0-7923-6928-9, May 2001

Volume 259: ***The Dynamic Sun,*** edited by Arnold Hanslmeier, Mauro Messerotti, Astrid Veronig
Hardbound, ISBN 0-7923-6915-7, May 2001

Volume 258: ***Electrohydrodynamics in Dusty and Dirty Plasmas- Gravito-Electrodynamics and EHD,*** by Hiroshi Kikuchi
Hardbound, ISBN 0-7923-6822-3, June 2001

Volume 257: ***Stellar Pulsation - Nonlinear Studies,*** edited by Mine Takeuti, Dimitar D. Sasselov
Hardbound, ISBN 0-7923-6818-5, March 2001

Volume 256: ***Organizations and Strategies in Astronomy,*** edited by André Heck
Hardbound, ISBN 0-7923-6671-9, November 2000

Volume 255: ***The Evolution of the Milky Way- Stars versus Clusters,*** edited by Francesca Matteucci, Franco Giovannelli
Hardbound, ISBN 0-7923-6679-4, January 2001

Volume 254: ***Stellar Astrophysics***, edited by K.S. Cheng, Hoi Fung Chau, Kwing Lam Chan, Kam Ching Leung
Hardbound, ISBN 0-7923-6659-X, November 2000

Volume 253: ***The Chemical Evolution of the Galaxy***, by Francesca Matteucci
Paperback, ISBN 1-4020-1652-2, October 2003
Hardbound, ISBN 0-7923-6552-6, June 2001

Volume 252: ***Optical Detectors for Astronomy II***, edited by Paola Amico, James W. Beletic
Hardbound, ISBN 0-7923-6536-4, December 2000

Volume 251: ***Cosmic Plasma Physics***, by Boris V. Somov
Hardbound, ISBN 0-7923-6512-7, September 2000

Volume 250: ***Information Handling in Astronomy***, edited by André Heck
Hardbound, ISBN 0-7923-6494-5, October 2000

Volume 249: ***The Neutral Upper Atmosphere***, by S.N. Ghosh
Hardbound, ISBN 0-7923-6434-1, July 2002

Volume 247: ***Large Scale Structure Formation***, edited by Reza Mansouri, Robert Brandenberger
Hardbound, ISBN 0-7923-6411-2, August 2000

Volume 246: ***The Legacy of J.C. Kapteyn***, edited by Piet C. van der Kruit, Klaas van Berkel
Paperback, ISBN 1-4020-0374-9, November 2001
Hardbound, ISBN 0-7923-6393-0, August 2000

Volume 245: ***Waves in Dusty Space Plasmas,*** by Frank Verheest
Paperback, ISBN 1-4020-0373-0, November 2001
Hardbound, ISBN 0-7923-6232-2, April 2000

Volume 244: ***The Universe***, edited by Naresh Dadhich, Ajit Kembhavi
Hardbound, ISBN 0-7923-6210-1, August 2000

Volume 243: ***Solar Polarization***, edited by K.N. Nagendra, Jan Olof Stenflo
Hardbound, ISBN 0-7923-5814-7, July 1999

Volume 242: ***Cosmic Perspectives in Space Physics***, by Sukumar Biswas
Hardbound, ISBN 0-7923-5813-9, June 2000

Volume 241: ***Millimeter-Wave Astronomy: Molecular Chemistry & Physics in Space,*** edited by W.F. Wall, Alberto Carramiñana, Luis Carrasco, P.F. Goldsmith
Hardbound, ISBN 0-7923-5581-4, May 1999

Volume 240: ***Numerical Astrophysics,*** edited by Shoken M. Miyama, Kohji Tomisaka,Tomoyuki Hanawa
Hardbound, ISBN 0-7923-5566-0, March 1999

Volume 239: ***Motions in the Solar Atmosphere,*** edited by Arnold Hanslmeier, Mauro Messerotti
Hardbound, ISBN 0-7923-5507-5, February 1999

Volume 238: ***Substorms-4,*** edited by S. Kokubun, Y. Kamide
Hardbound, ISBN 0-7923-5465-6, March 1999

For further information about this book series we refer you to the following web site: www.springeronline.com

To contact the Publishing Editor for new book proposals:
Dr. Harry (J.J.) Blom: harry.blom@springer-sbm.com

GPSR Compliance

The European Union's (EU) General Product Safety Regulation (GPSR) is a set of rules that requires consumer products to be safe and our obligations to ensure this.

If you have any concerns about our products, you can contact us on ProductSafety@springernature.com

In case Publisher is established outside the EU, the EU authorized representative is:

Springer Nature Customer Service Center GmbH
Europaplatz 3
69115 Heidelberg, Germany

Batch number: 10607214

Printed by Printforce, the Netherlands